Production Technology of Fruit Crops

Production Technology of Fruit Crops

Edited by:

Mohan Balasaheb Patil

Delve
PUBLISHING

www.delvepublishing.com

Production Technology of Fruit Crops

Mohan Balasaheb Patil

Delve Publishing
224 Shoreacres Road
Burlington, ON L7L 2H2
Canada
www.delvepublishing.com
Email: orders@arclereducation.com

ISBN: 978-1-77469-018-5 (Hardcover)

ABOUT THE EDITOR

Dr Mohan Balasaheb Patil is presently working as Professor and In charge of fruit research station Aurangabad. He did his Bsc Agri. from Marathwada Agril. University Parbhani in 1986. Msc (Ag) Horti. In 1988 while, Ph. D in 2004 from the same university. He has started his carrier in 1994 as Asistant Prof. promoted as Associat Prof. in 2007 & Professor in2017. He has established the Sweet Orange Research Station at Badnapur. Published 25 research papers in national and international journals. Published more than 100 popular articles. He reeived Agrocare Award for the best extension activities. He has authored and published two books.

TABLE OF CONTENTS

LIST OF FIGURES

PREFACE

Fruit crop production plays a central role in the success of any country. It is typically believed that people's living standards can be determined by their fruit production and consumption per capita. Compared with other field crops, fruit crops have a higher yield per unit area. For instance, a wheat crop yields around 12-15 tons per hectare of land of two crops a year. This volume mentions that fruit trees are the most productive if they are carefully bred using the appropriate technology.

The topics discussed include chapter 1: introduction to fruit crop production, chapter 2: classification of fruits, chapter 3: importance of fruit production, chapter 4: planting systems and transplanting of fruit trees, chapter 5: orchard establishment and management, chapter 6: use of growth regulators in fruit production, chapter 7: fruit training and pruning, chapter 8: integrated water and nutrient management of fruit crops, chapter 9: the use of greenhouses for fruit crop production, and chapter 10: organic production of fruit crops.

All fruits need well-drained soil with decent water retention capacity. Another important factor for successful horticulture is to obtain the necessary inputs, such as planting materials, regular high-quality water supply, recyclable soil and pest management. The shrubs and weeds must be removed before planting, deep plowing and leveling the ground. Automation technology for weed management can be applied for this purpose. Machine vision technology and data processors have been made to enable commercial machines to identify crop line patterns and manage automated equipment that carry out tasks such as removing weeds in the line and thinning crops to the desired stand.

The volume also points out that the most common mistake people do is to plant a single fruit tree. It is not just important to plant good-tasting and high-yielding varieties, but also to understand some pollinator requirements when choosing fruit trees. If the planted fruit varieties have self-incompatibility, the flowers and fruits can bear fruit, but will eventually fall. Most fruit trees develop best when two or more different breeds are planted. The species should bloom approximately at the same period and have compatible pollen. Growers must check whether the varieties available in the fruit tree catalog can pollinate one another. If space is limited, looking for fruits and varieties that can reproduce themselves is recommended.

Intercropping is another technology that is gaining popularity among fruit farmers. This is the process of growing cash crops in the space between fruit trees in the initial years. They can also be used as agricultural crops and gain land benefits through intercropping, irrigation and fertilization. The water requirements of intercropping must not conflict with the water requirements of the primary fruit trees. Compared with millet, vegetables make up the best intercrops. However, regardless of interplanting, the commercial plants must be kept away from the main fruit trees and irrigated separately. When trees occupy the whole orchard space, interplanting should be stopped. Thereafter, only green fertilizers or cover crops should be applied. The volume suggests that to determine whether a fruit variety can thrive in a given region, fruit production professionalsmust be consulted.

Chapter 1

Introduction to Fruit Crop Production

CONTENTS

1.1 INTRODUCTION

Growth of fruits can be an enjoyable experience, a demanding sport, and a relaxing experience. There is a reason for growing fruits. Fruits add a lovely color and variety to your greenhouse. In a comparatively small area, an immense amount of fruits can be produced. When they're completely ready, shopping fruits are always picked, delivered and sold, and stock choices are normally better, but not necessarily the best. The new fruit from the garden also brings the family sweet, nutritious and healthy food, offers a good source of vitamins, nutrients, biomass and fiber. You will taste the fruit is much better than the fruit you buy in the grocery store, as an added reward. Long-term commitment is the production of an oak grove that requires critical planning, appropriate location and site selection, planting methods and planting distances for optimal yield, as well as varieties and nurseries. Crop cultivation in containers in your garden can also be appreciated if you have no storage. For a couple of years, any fruit tree can be grown in a pot and then transplanted. You can also prefer a dwarf form that is ideal for living in a pot. You will harvest dessert-quality items if you carefully select the shape and variety of the fruit before planting, from early spring to fall. Many people provide a natural landscape for plants and are aesthetically designed more esthetically in the greenhouse, making it more difficult to restore. How do you pick plants for a fruit orchard? "There are so numerous fruit and vase trees. "There are plants with an esthetic appearance, and they are also ideal for edible landscaping.

1.2 PRE-PLANTING

Site Selection

Fruit plants are most successful when carefully matched with the correct setting and planting site. To get field workers quickly and also to profit from the expertise of other fruit growers and to sell goods through cooperative fruit grower associations, the place should be close to the market and well-established fruit growing area. With good water holding capacity, all fruits require well-drained soil. Access to the necessary services, such as planting materials, regular supply of high-quality water, environmentally sustainable soil and pest control, live fences and nearby inexpensive credit or gardening income, is another crucial factor for successful cultivation. Choose a place free of frost patches, wet spots, cyclones, frosts, hailstorms, high hot winds, strong winds and enough sunlight for the best fruit production. Also in partial shade, gooseberries and currants perform well enough, while other fruits require full sunlight for at least six hours a day, and hopefully more. But if your key goals are development and ease of management, devote various areas to a selection of fruit trees and small fruits. Strong road or rail transport networks must be available (FAO, 2020).

Before planting, the shrubs and weeds should be cleared the site thoroughly ploughed and graded. Based on the topography of the land, the land should be divided into terraces in the hilly areas and the terrace leveling should be carried out. Terracing is shielding the earth from erosion. If the soil is poor, it will be better to cultivate and plough a green manure crop to strengthen its physical and chemical conditions before planting. Until a year or two of planting fruit plants, growing marigolds, Sudan grass or some mustards (oilseed rape) can help to contain certain parasitic nematodes that are often of concern in certain soils. Plants such as thorny Agave, *Prosopis juliflora*, *Pithecolobium dulce*, Thevetia are resistant to drought, easy to replicate from seeds, quick to grow, have dense leaves and intense pruning, and would serve as good live fencing if closely planted in three rows. In areas where strong winds prevail, planting two rows of tall, fast-growing, hardy, dry-resistant and mechanically strong trees, such as *Casuarina equisetifolia*, *Pterospermum acerifolium*, *Polyalthia longifolia*, *Eucalyptus globulus*, *Grevillea robusta*, *Azadirachta indica*, etc., is important to act as windbreaks. To reduce competition between the roots of windbreaks and fruit trees for moisture and nutrients, it is advisable to dig a trench 90 cm deep at a distance of 3 m from the windbreak trees.

Because it is difficult to repair nutrient deficiencies and adjust soil pH after planting to see whether you need to add lime and nutrients, it is necessary to test the soil before planting. Collect subsamples containing a representative sample of the site from multiple locations. Pass soil samples to a professional soil research facility situated in the vicinity. The pH, the capacity for exchange of action and the concentrations of various nutrients and soluble salts present will be reported in the soil test. The soil pH is a measure of soil acidity. A pH of 7.0 is neutral; a pH of acid soil is below 7.0 and a pH of greater than 7.0 suggests that the soil is alkaline. The optimum pH range for most fruit crops is 6.0-6.5. Only blueberry grows well in acidic soils (pH 4.0-5.5). You should use lime if the soil test indicates that the soil is too acidic. Lime can increase the pH of the soil to provide the necessary calcium or magnesium. However, do not add lime until a soil test recommends. Supply of phosphorus, boron, zinc, iron and manganese can be decreased by excessively high soil pH. Reducing soil pH is more complicated. If the soil has high carbonate content, the application of sulfur to the soil can decrease soil pH. If you want to grow blueberries in very alkaline soils, it is safe to remove and substitute peat moss or another low pH organic matter for the soil in the root area. Soluble salts refer to soil electrolyte compounds that dissolve in soil water and are referred to as the degree of salinity. Soluble salts at high levels can reduce plant water absorption, reduce root growth, cause root tip burns, and generally reduce plant growth and fruit yield. Soluble salts compete with plants over soil water (FAO, 2020).

The resistance of plants to soil salt varies. Salt tolerances are established for single fruit crops. Fertilizer discharges or fertilizer closures to fertilizer roots can cause soluble plant salt problems. Adding organic matter or planting cover crops for one year before planting and correcting soil problems before planting. If immediate planting is necessary, cover with sphagnum moss, moist cloth or other material to keep it from drying out and store at a temperature above zero. Keep the roots moist while the water is poured, if possible. Lighter plants would be 'healed' by planting a limited amount of loose soil as soon as possible.

Figure 1.1: Fruit farming

Source: https://www.richfarmkenya.com/2017/03/passion-fruit-farming-presents-income.html

1.3 REMOVE FROM THE ORCHARDS

Any design policy should be designed to have the greatest number of trees per hectare, ample space for proper treetop growth and encourage the activities of planted orchards. Architecture can be divided into two large groups, i.e. the following are the most common interface schemes:

Square method

In this arrangement, the trees are rooted, no matter how deep each square can be planted. This is the most commonly practiced type that can be defined in two directions and can be intercropped and cultivated. The central position between four trees is useful for short-lived fillers.

Rectangle method

The trees on either side of the rectangle are part of this scheme. There is more than the distance between two rows of trees in a row. A greater variety of fields makes for fundamental mechanical and intercultural practice.

System Hexagonal method

The trees are rooted in the equilateral triangle of each area. The six trees thus form a hexagon, the seventh tree in the centre. This arrangement is also known as the "septula" in that the seventh tree is situated in the middle hexagon. The strategy has good separation, but cultural activity is difficult to define and organize. The scheme includes 15% more trees than the rectangle.

The Quincunx or Diagonals method

This is a square tactic, so it's in the centre of the square. It can accommodate double the number of seeds, but the spacing would not be the same. The picked central filler tree should be short-lived to compensate for when the main trees come into being after a few years. This scheme would be followed if the gap between permanent trees is 10 m.

Method of the triangle

The triangular form is based on the concept of a single triangle. The distance between the two adjacent trees, perpendicular to each other, is equal to the distance between the two rows. There is more space in each tree than in the square system, and thus fewer trees per hectare than the square system is allowed.

Scheme of the contour

Typically followed by hills with undulated topography to minimize soil erosion and retain soil moisture. Plants are planted along the slope of this device. The line of contours is formed and graded in such a way that drainage canal water flows continually, thereby providing time to enter the surface without erosion. Planting on a flat strip of soil, built on a contour, is carried out using a terrace operation. Terraced fields grow in stages above each other, helping to produce more soil and prevent soil erosion. The distance of the contour terrace depends on the configuration of the slope. The seed distance may not be constant during the contour process.

1.4 PLANTING DISTANCE

The distance to planting in different fruit crops is determined based on tree formation, tree form, rainfall pattern, soil type and soil fertility, rootstock vigour and planting and training methods. If the spacing was too broad, the output per unit area would be slightly lower. Planting trees more closely and supplying the requisite water and food materials are beneficial. Plants grow tall as trees are too close to prune, spray and grass. The plants are rising tall. Water and nutrients are competitive root and tree yields less or produce thin, badly coloured berries. Near planting results in higher yields per unit area in the early stages of the tree, but less lately.

The hexagonal planting style is 15 percent larger, whereas the triangular shape is 11 percent smaller. When planting in a square or rectangular pattern, it is very easy to calculate the number of trees per hectare, and are rendered by dividing the total area by the area occupied by each tree (a x a in the square system or l x b in the rectangular system) (Valpuesta, 2002).

1.5 HIGH-DENSITY PLANTING UNIT

Innovations have been developed to allow effective use of horizontal and vertical space in mango, citrus, papaya and guava. Instead of being similar to the recommended fruit tree planting, it is often referred to as 'High-Density Plantings' or HDP, with some special techniques with the primary objective of achieving maximum yield per unit area without sacrificing efficiency. Apples were first introduced in Europe in the 1960s, and most apple orchards are now being produced using this method in Europe, America, Australia and New Zealand. This strategy considers four apple crops: low hDP, moderate hDP (250-500 trees/ha), large hDP (500-1250 trees/ha) and strong haDP (>1250 trees/ha). Also in apple trees with a population of 20,000 trees per hectare, a mega high-density planting scheme has recently been implemented. Early crushing, longer-term yields, lower labor costs and increased fruit consistency are the benefits of HDP. The planting of apple trees (saying 70,000 trees/ha) is also more closely carried out in some orchards, also referred to as 'meadow orchards.' In traditional planting systems, the average yields found for apples are roughly 5.0 t/ha and 140.0 t/ha for HDP. HDP trees have the largest number of fruit branches and structural branches. The trees are typically formed by a central leader surrounded by almost horizontal fruit branches. These branches must be pruned in such a way that the shadows on any branch on the other branches are negligible. The output of the HDP varies with the size of the tree. Dwarfing and intermediate rootstocks, such as MM 106, MM 109, and MM 111, can be used in oranges, Quince A, Adam, and Quince C in pears. The use of scissors of the spur type also causes dwarfness through teaching and tailoring techniques. Apple trees are best matched to HDP systems and are taught under spindle bush, dwarf pyramid and cord systems. Production inhibitors, such as diaminoazide, ethephony, chlormaquat and paclobutrazal, are also used to limit shoot growth by between 30 and 0 percent. This will lead to a better flowering in the coming years. The branches should be connected to a 45° angle from the main stem, a number of the normal methods used to regulate the height of the tree. The efficacy of HDP relies on the correct selection of planting methods. Rectangular planting is usually practiced for single, double and three-row trees. For single-row planting, the distance between rows is small and the distance between rows wide (4 m x2m). One of the two rows (4m +2m x2m) is divided in a double row, and each of the three rows (4m +2m x2m) is split by a narrower row into three rows. 10 to 15 rows of beds are extensively planted within the meadow orchard system (30x45 cm) and 2.5 meters wide streets are separated between the beds. This scheme is sometimes referred to as the bed process (FAO, 2020).

It depends, in part, on how far away you are from the fruit trees in space to protect them. Trees are ideally planted between 12 and 18 feet apart last year. Trees that are less frequently pruned grow bigger and require more space. Highly pruned trees need a lot less area, often just 8 to 10 feet, several times a year. The growth conditions also decide how large trees grow and how much space they need. In very fertile, damp soils, trees grow faster, while rocky, dry soils trees grow slowly.

Figure 1.2: High density planting system and required equipments

Source: https://www.kisankraft.com/high-density-planting-in-fruit-crops/

1.6 SET OF VARIETY

Any form of fruit tree, including cultivars, has its climate change and limits (variety). Stone berries are more suitable to tropical climates such as banana, pineapple and papaya, such as pine nuts, sweet and raspberryChoose the kind of fruit that can do best if your place is warm and dry. In any case, the mean temperature of the winter air would be below the survival limits of the crop if the fruit were grown outside its atmosphere and the flowers would be killed by frost during the season. The average estimated time is less than 7 °C a year. Chilling is named after a variety of cooler fruit trees and nut trees. Fruit trees can grow well without a sufficient cooling period but produce little fruit. Find out about the usual cooling cycles of your area and grow trees that suit your cooling needs. Fruit crops can differ from time to time for cooling, e.g. on low, medium or high chilled plants, e.g. between 300 and 1,200 hours of cooling. The most common mistake ever made is to plant a single fruit tree. In addition to the delicious, successful taste for the growers, it is important to know a little about the need to select a fruit tree for the pollinators. Flowers and fruits, though, can eventually decrease if the fruit of the seed is separated. A variety of fruit trees survive well when at least two different species are planted. The species evolves at around the same time as the pollen is compatible. View catalogues of fruit trees that pollinate one another. If your space is limited, look for fruit and varieties that are self-sufficient. To find out whether a variety of fruit is growing in your area, call the right man. Find ways to adjust the microclimate to increase your chances of success if you wish to grow fruit trees from another climate region. Grapes first require dry conditions during the ripening phase of the fruit. In the season, consider planting grapevines to ascend the porch below the eave to shield the fruit from too much rain (Md Asaduzzaman & Toshiki Asao, 2019).

1.7 PREPARATION

Do not sow or anticipate fine root seeds, low leaves and poor fruits in shade areas. When fully cultivated between the plantation and its houses, roots, power lines and other obstacles, the tree leaves stray to fill its space. It depends on the fruit, the temperature and the environment of the crop. A variety of tropical and subtropical plants, including bananas, pineapples, sapota or guava, are cultivated in the spring and rainy seasons, while in the winter, temperate cabbage plants are grown. When trees have been planted in early rainy season, they can grow quickly and vigorously. It is necessary to plant or weaken trees seriously before work starts, and the next warm weather will last. Trees are planted at a level with a central scale and a hole at each end of the planting board, typically 15 meters tall, 10 cm long and 2.5 cm wide. This platform is organized to match the tree with the central notch. Each of the final holes has one of the two small stakes. The board and tree marker are straight without twisting the last stakes. Dig a pit that is wide enough to cover the whole root system of the fruit plant with a diameter of about twice and a depth of two feet to cloak the roots properly. Place a mixture of gravel and yard dung in the pit (1: 1: 1). A minimum quantity of insecticide shall be used to track certain ground-born insects before planting. Run the furrows and let them go and have a rest. Dip the roots of the plant into a fungicide before planting. Disperse roots in a pit in a loose state to make sure they are not stuck or crowded. Cut them thoroughly if the roots are too long or too difficult to enter the distance easily. "If you plant a budded or grafted plant, care must be taken to see that the graft union is 2-3" above the ground floor. The grafting process is called 'Scion' or the true variety and the moisture-collecting rootstock and the nutrients are anchored in the grains in the grafting union and are characterized by their dwarfing quantity (NIIR Board, 2005).

Education shows a comparatively new approach in which the growth of the tree is directed towards a desired form and structure. To be more efficient, fruit plants have to be formed in some way. There are several educational programs available. Some fruit trees are trained on one trunk and others are trained on the Kniffin, vertical trellis, overhead trunk, middle or open centre trunk, and so on. Lowly fit fruit trees typically exhibit very steep angles of the branches, which create a significant rupture of the branch under tremendous fruit volume. This greatly decreases tree productivity and significantly reduces tree longevity. The tree canopy may also be opened to increase light visibility during adequate tree planning. For most leafy trees, flower buds will be formed during the previous summer for this season's harvest. Floral buds and an optimal fruit set, taste and texture require light penetration. While a mature tree can be planted in full sun, a dense canopy does not allow enough light to reach the canopy between 12 and 18 inches. The opening of the tree canopy also allows ample movement of air through the tree that facilitates quick drying to reduce the visibility of pathogens and helps the pesticide to penetrate deeper. Also, a well-formed fruit tree is aesthetically pleasing, whether in a landscaped garden or a home garden. Planning is adequate if regular pruning is done properly (NIIR Board, 2005).

1.8 TAPPING ACTIVITIES

Tapping is a deliberate removal of plant sections to ensure equilibrium between foliage and sustainable growth. It has a huge effect on plant fertility. In Deciduous fruit, it is necessary for plumage to maintain its fruiting speed, i.e. that there is no or little plumage of peaches; that annual plumage is vital in ber; that there is annual plumage in northern India in grapes, two in Maharashtra and three in Tamil Nadu. Pruning is normally achieved in evergreen fruits, but in certain species, such as mango and guava, it is helpful. Another error is the height of the leaves. Taping trees of the right size keep them healthy, which is vital for optimum tree survival and fruit production. Keep an adequate number of healthy leaves to process successful fruit before cutting. Good quality fruit requires around 30 to 40 pleasant, large leaves. The best time to pick most trees after the summer harvest to reduce their growth is after harvesting. Winter pruning can only take off suckers, deadwood and fallen or crossed branches to loosen the leaves of the tree. Cuts should be made at an angle to avoid the absorption of moisture to settle rot and illness. To deter pathogens from joining, all cuttings should be covered with Bordeaux paste (Ladaniya, 2010).

1.9 FRUIT PLANTS THAT ARE FERTILIZING

Figure 1.3: Fruit Tree Fertilizer Boosters | Bio Wash Fertilizer Boosters

Source: https://fertilizerboosters.com/fruit-tree-fertilizer-boosters/

Do not use fertilizer when planting a fruit tree. It's going to burn the young roots with the fertilizer and cause more harm. The general recommendation is that 100-150 grams or 10-10-10 fertilizer equivalent should be applied 7-10 days after planting and 40 days after planting. Spread the fertilizer uniformly and work 8 to 12 inches away from the trunk above a few inches of mud. Plants should be well-watered after fertilization. In the second or third year of planting, it would be advantageous to add 250–300 grams from 10–10–10 in March and again in May. A double quantity of 10-10-10 or similar fertilizer can be used in light sandy soils where phosphorus and potassium may be low. However, it should not over-fertilize because the fruit yield and quality can lead to too much vegetative growth or cause root and/or harm. Avoid using salt-containing fertilizers as this would impede plant growth. It helps to provide organic matter to a range of plant nutrients; it increases soil structure and texture by adding looser texture to the firm, hard-to-work soils and retains more water in sandy soils. Once soil analysis indicates that the organic matter is less than 2%, organic matter can be included in the soil in the top 6 inches before planting. A variety of items are eligible for selection, including leaf manure, vermicompost, cow manure, organic compost, etc. The need for organic fertilizers is not only helped by biologically sound products such as Trichoderm, Vesicular Arbuscular Mycorrhiza (VAM), Phosphorus Solubilizing Bacteria (PSB), etc. Panchakavya, an organic substance made from portions of fresh milk, curd, ghee, urine and dung, can also be used as a foliar spray on fruit plants to improve plant growth and disease resistance.

The use of fertilizer for maturation should rely on the growth and vigor of the tree. The fertilizer would disperse around the drop line, while the roots of the tree feeder would be distributed around the drop line. During the first season, the quantity of other nutrients needed for fruit crops is best shown by a leaf inspection. Shortly after they are old enough to survive, fruit trees begin to bear fruit. Pollination and berries only blossom once a year, with some fruit trees and varieties flowering two to three times a year based on conditions. There are other fruit trees, too. However, you affect the ability to bear fruit regardless of its health, fruit preferences, and cultural customs. Adequate pollination for fruit yield is important. Any fruit can be damaged or decreased under unfavourable conditions. Separate male and female flowers are found on the same branch, some trees like pecans. Where the male pollen is lowered before the female flower is sufficient, the fruit aggregate is a problem. No other category is suited to any sort of fruit trees. Pistachio includes male trees and female fruit trees. At least one male tree must be planted for each of the eight female pistachios to be successfully planted. Each apple tree is infertile to itself. Plant at least two distinct varieties together. 'Golden Sweet,' a self-fertile variety, and 'Jonathan' are the most common pollinators, which also supply a significant amount of fruit trees in one year, slightly later in the next year. The spring-flowers of the most robust fruit trees are produced in the middle of spring or summer. This is called 'biennial bears.' So the trees of next year's harvest will be seriously discouraged or killed by a very strong year's crop. It is difficult to alter or fix the two-year bearing found in mangoes, apples, etc (T. Radha & Lila Mathew, 2007).

Why is the fruit in the garden so bad?

Fertility is a state in which the plant produces flowers, the fruit grows and matures. Infertility is a condition in which a plant cannot bloom, develop or mature. This is a big challenge in many fruit trees, and both external and internal variables need to be weighed to obtain an acceptable production level. External variables are factors that influence the life cycle of a plant, e.g. before or after flowering, and include the environment. Climate: temperatures, humidity requires, gloomy weather, wind and frost; age and strength of plants; rootstock, soil and water disturbances; pesticide supply, plasters, pests and diseases; chemical spraying during flora.

Plant temperature conditions are very easy for growing, flowering, and fruiting. The effect on seed fertility, embryonic development, and growth are widespread. Depending on the form of fruit, the maximum temperature varies from 22-28 °C. Optimal temperature differential impacts the behaviour of the flora; the initiation and pollination of the fluids can be affected day-night. Flowers are adversely affected and extreme temperature fluctuations (high or low) hinder pollination and fertilization. Temperatures over 32 °C cause the stigma surface to die and the embryo bag to decline faster.

Flowering precipitation hurts in a variety of ways, washing anthers, stigmatizing pollen, decreasing the viability of pollen, reducing the activity of bee pollination, reducing fruit or no fruit. Strong rain may lead to a substantial drop in fruit. Diseases such as powdery mildew, typically arising directly after cloudy weather, delaying pollen dehiscence, disrupting the germination and development of pollen, and having a destructive effect on flowers and fruit production, are all rendered more effective by cloudy weather.

The wind is the most effective pollen transfer agent for pistil transmission. This is critical in both the insect (entomophilic) and the wind (anemophilic). Bees and other pollen-holding insects work more successfully in a silent environment. Equal wind speed encourages pollination by shifting pollen, which is necessary for the development of fruit. Strong wind speed prevents the movements of insects and affects pollination during the flowering season. Strong and drying winds hurt the fruit environment (T. Radha & Lila Mathew, 2007).

Air temperatures rising to 0°C or below 0°C are the atmospheric conditions for frost. Water vapour transforms into ice and sinks on the top of the plant, causing irreversible damage to the plant. Blossoms or sexual parts of flowers can be damaged that renders plants ineffective. Many plants are young for a long time and only bear after some maturity. New and strong trees are no longer fruitful.

The mechanical and physiological aspects influence the rootstock on the scion. Grafted crops, for example, grow early plants and hinder fertility relative to seedlings. Troyer citrus rootstock of citrus fruit hastens towards fruit maturity. SohSarkar rootstock in citrus hastens to maturity.

Disrupted soil-plant water links can lead to flowers and fruit being abscessed. Water shortages are especially harmful in the early stages of fruit development. Humidity stress triggers the disturbance of the carbon-nitrogen ratio (C/N ratio) and other physiological activities that lead to a decrease in fruit. The best performance is a sound source of water. Flowers and fruit like apples and olives are affected by excess water.

Inadequate pruning, unnecessarily harsh pruning, and pruning can lead to poor flowers and fruit picking at the wrong period.

The number of fruits in flowers is vulnerable to attack by several rodents, mice, and viruses. Insects, domestic and honey bees are useful for fruiting and insalubrious (mealy worm, anthracnose, and mango hopper). Fruit collection substantially increases the execution of successful pre-attack mitigation techniques. Useful insects pollinate; cultural activities are planned to deter harmful insects. Noxious insects, such as a mealybug, sweep flowers and grow fruit and cause flowers and fruit to drop. The threshold can also be lowered with appropriate preventive and timely control measures (T. Radha & Lila Mathew, 2007).

Evolutionary trends (sexuality) are internal factors; monoecious, dioecious and pseudohermaphroditism in the area of sexual distribution; biological fluid distinctions (flotation and formation); heterostyled abortion, dichogamy, pistil or ovule, pollen impotence and plant nutrients (C/N ratio).

Monoecious (Unisexual): the flowers are imperfect; the wind is often pollinated on the same herb, with the same inflorescence or with various inflorescences, stagnant or pistillate. A crossover in spatial isolation could be possible but would be less likely to be assured. It is common in temperate areas. Examples include chestnut, hazelnut, walnut, jackfruit, pistachio, and date palm, papaya, coconut and isca nut.

Dioecious (Unisexual): male and female seeds of various plants. For proper pollination, fertilization, fruit setting and seed development, several males must be planted in the vicinity of female plants. Otherwise, poor fruiting is going to happen. It's usually found in the tropics. For example, papaya, palm date, raisins. A variety of Japanese persimmon varieties are monoecious.

Blossoms in both male and female portions but not functionally monosexual (either stem or pistil, male or female), e.g. grape, ink, granite, persimmon. Pseudoshermaphroditism is morphologically perfect.

Structural diversity: certain minor features of normal flowers are either in floral type or in functions that impede self-pollination and make cross-pollination a rule; lack of cross-pollination contributes to incompatibility and infertility of oneself or cross-pollination. E.g., heterostyled, dichogamy, pollen impotence, pistil and ovule abortion (Hancock, 2008).

Heterostyled: the same plant has a different type of flower. Heterostyled, for instance: the appearance of theme-long vector flowers (common in Prunus fruit). Distyle: short style and long filaments (a type of thrum); long style, short filaments (a type of pin); (compatible with pin type and thrum type or vice versa). Cashew, almond, freshness and sapota, e.g. long, medium or short pomegranate, litchi.

Dialgamy: stamens and pistils in hermaphrodite plants grow at different stages, and the receptivity does not reach the viability/maturity of the pollen, preventing the self-pollination of large monoecious flowers.

Protandry (stigmatic receptivity is not associated with pollen viability and maturity): mature stamina and anther are shed before the pistil can tolerate pollen; more generally than protogyny, usual in pollinated plants, it ensures temporary separation. Walnuts, cocoa, cuddly apple, sapota, passion fruit, for example, are all part of this.

Protogyny: stigmas are receptive to the release of pollen by the anthers. The protogynous plants with a large number of dioecious plants are also monoecious. E.g., pineapple, avocado, chestnut or pistachios.

Abortiful flowers impotence: the reflexed stammers, or the complete pollen produced by perfect flowers, are thin or unviable (for example, European grape varieties), and do not result in mature fixed fruit. Abortive flowers are more common in plants with an indeterminate inflorescence. High pollen viability is needed to ensure a good fruit set and yield.

Degenerated or abortive ovules impotence: any factor that hinders the mechanism of fertilization results in fruitlessness or sterility. Full: no flower or sex organs are formed, or not fully formed; partial: abortive only stamens or pistils that often look natural; ovules abortion: e.g. mango; in male and female, sterility in several ovules and anthers is gained.

Genetic incompatibility means that a plant does not produce a zygote out of its pollen. There are two types of sporophytic and gametophytic. There would be no success in pollination if the pollen had the same allele as the stigma.

Vegetative growth, poor pistil quality and powerless pollen are degraded by their low nutritional status. Excessive N fertilization results in excessive plant growth which disrupts the C-N ratio at the expense of flowering. The CC/NNN reflects the highest C/N ratio.

A variety of young fruits are usual in the offspring and early summer seasons. This natural dilution is often referred to as 'June Fall.' A lot of fruit is supposed to stay on the tree in other cultivars. If it's still really thin, the excess fruit must still be collected manually. The thinning of the plant prevents the rupture of thebranch. It increases the size of the seed, improves the consistency and texture of the residual fruit and prevents decay of the tree (Hancock, 2008).

The management of viruses and insects is a big concern once the trees have started to flower. Take a daily peek at the trees to see if there are fresh victims. Insects such as aphids, leaf rollers, mites, moths, slugs and maggots all eat fruit trees. Other than flies, molds, mildew, bruises, scabs and brown-red can also cause issues with fruit trees. Successful management of insects and viruses is driven by sanitation. Extract and destroy infested or contaminated vines, rabbits and field fruits to discourage infection. Sleeping oil tracks the scale of the psylla, the mites and the pear added before the buds are punched. Many soft-bodied insect species are contaminated with insecticidal soaps. Copper and sulfur-containing products can control fungal and bacterial diseases. Since insect and virus infestations can be extremely contagious, any problems need to be addressed quickly to prevent infecting neighbouring plants.

Pest management has been successfully introduced to many crops with *Bacillus thuringiensis*, a disease agent capable of targeting several lepidopteran insects. The spread of sterilized insects with pheromone traps or stick traps is another form of pest control. Many plant pests and diseases are governed by botany. A variety of insects, bacteria, fungi and viruses may be controlled by neem, dhatura, Calotropis, Pongamia. These pesticides are cost-efficient, environmentally friendly, and ready to prepare for multiple pests and diseases and are effective in their treatment. In many fruit trees, birds are by far the main pests. Perhaps the safest surveillance device is a plastic net that protects trees with wireframes. The nets may include smaller trees. Using a tight mesh net to minimize the risk that birds get caught on the net. Individual fruit can also be wrapped to discourage injury to birds, damage to insects and attacks by butter or fabric bags. Aluminum tins, which are suspended by a cord or wire over the trees, are turned into the open, keeping the birds away.

Fruit harvesting is either an individual or in clusters, depending on the shape or type. Few varieties can be easily shaken off when they are ready, while others must be hand-picked. A ton of fruit, while still firm in colour, should be picked, but rather green. They turn brown and fluffy on the branch if left to ripen. Color, firmness and taste help determine if the fruit is ready to be picked (Hancock, 2008).

There are few suggestions online.

- Look for a transition from green to yellow skin tones;
- Or the colour of the brown coat;
- It's raw, fleshy, pulpy to crisp and juicy in texture.
- Fresh herbal scent of orange and ripe fruit

Take the fruit out of the tree, as you have selected, by going up. In shelf time, the fruit is comparatively short. Place the fruit in the refrigerator or other cold storage for a significant amount of time. A tree should produce enough fruit to satisfy the nutritional needs of a typical Indian household.

1.10 FROST, FROST PROTECTION AND WINTER DESTRUCTION

Since farmers started growing, the damage caused by frost in fruit crops has been a problem. Frozen loss is a global concern, as the freezing temperature of the night would lead to total crop failure if crop production is controlled in all its aspects. In Bagdonaset, cold air weights in subtropical climates are actively changing and frost periods are shorter in length and occur more often in temperate areas. In the tropical world, except at higher altitudes, there is typically little too little freezing impact. Damage also happens during the winter in subtropical crops, often in the spring, but also in the fall in lagoons and nut trees. The date of field planting, microclimate, variety and other security measures for a region vulnerable to frost must be changed. Before 1835, however, farmers had opted not to plant oranges in South Carolina, Northern Florida and Georgia due to losses due to frost damage. Frost damage has become more sporadic and often causes many severe injuries.

In December 1990 and 1998, the California citrus industry sustained two significant losses of frost leading to moderate damage to serious damage in some areas. Fruit plants had a good economic effect in the freeze of 1990 compared to the freeze of 1998. In 1907, in Grand Junction, Colorado, USA, for frozen protection, Scheu invented metal jar heaters (such as stack heaters or smudge pots). The heaters in the metal tub are made up of a heating oil-burning unit, later known as the HYLO orchard heater. Heaters have mostly been used in fruit trees to shield crops. However, in modern times, metal fire tanks have been used to retain heat for radiation and seed convection.

The use of heaters has been common practice in gardens all over the world for some time, but later in America, the use of orchard heaters has been outlawed, because the smoke has seriously polluted the air due to health and climate. Return stack heaters and clean propane fuel heaters are legal in many places today. Farmers continued to carry out their crop insurance after the freeze of 1990, and farmers with insurance sustained further losses in 1998. Since their hedges were more vulnerable to destruction or because they were protected, no attempt was made to enforce security techniques. Frost The term "frost" is defined as being at temperatures of 0 °C or below, at maximum altitudes of 1.25 and 2.0 m above ground level, the temperature of the air at an acceptable level of weather protection. The word frost is often misunderstood for the words freeze, but it has several interpretations (Zhang, 2017).

Frost is a situation in which water in plants may or may not be filled while 'freeze' is a therapeutic condition that ranges from liquid to ice in extracellular water, depending on several protective variables (e.g. super freezing and agglomeration of nuclear ice) (i.e. 35 freezes). Depending on the strength of the plant, freezing may or may not cause tissue damage (e.g. the solvent content of the cells). When a plant or tissue forms extracellular ice, the frost event becomes a freezing event. Freeze damage is caused by a drop in the vital value of the temperature, which leads to the death of plant cells.

Figure 1.4: Severe frost damage to orchards across northern Italy

Source: https://watchers.news/2020/03/28/severe-frost-damage-to-orchards-across-northern-italy/

Radiation/Congelation

Radiation freezes with a bright sky and a quiet wind speed of 5 mph (less than 8 km/h) contribute to a drop of less than 0°C during the day, low dew points and nocturnal air temperatures of less than 0 percent. The temperature of the dwarf point is the temperature achieved by cooling the air until the relative humidity is 100 percent and the water vapour level of the air is directly determined. Heat radiates in the radiative ice from the surface of the earth to the outer atmosphere, that is, the sky. Radiation Style Frost: often composed of two types. Frozen white: happens when the soil deposits ambient humidity and creates an ice sheet. It's less damaging than the black ice. Black frost: this kind of freezing happens when the temperature drops below 0°C, resulting in little to no development of ice crystals since the weather in the lower atmosphere is too dry. Ice crystals are formed according to the point of dew or frost. The black frost is worse than the frost of the skin (Zhang, 2017).

Advective Frost is characterized by intense, dry, cold and cold winds (450,000-3,000 ft.) and medium to strong winds, low humidity, no temperature change, and a dew point lower than zero in certain cases. This is partly because a mass of cool air enters the atmosphere and consumes the hot air that allows it to freeze. Temperatures frequently dip below the freezing point (0 °C) and remain there all day long.

Restricted Frost Advective Safety Efficiency

Where is the ice going? The soil heats and heats up in the orchard throughout the day. In the night, and to some degree, out of the fruit trees and the vines, the heat rises out of the earth. The freezing happens when the temperature rises to 0 °C on the ground.

Decrease in temperature depends mostly on the following factors:

- The amount of heat stored in the soil during the day.
- The amount of heat wasted by radiation during the night;
- The transfer of heat from the soil to the base or surface of the plant.
- Quality of air humidity. At night, when cloud cover and wind, radiation loss is the highest as it mixes clouds of ozone and forces out cooler air.

Water vapor has been turned into water (dew). The temperature of the air in contact with plants falls below the "dew point" of the soil at night, which allows condensation and dew to cause moisture. This is also the only way to transform steam into water (dew). This slows down the temperature decline. It falls as the temperature drops to 0°C and the heat is released as the dew transforms to ice and snow.

Frozen Security Approaches

- Passive protection or passive strategies could be used in expectation of the real danger of freezing. They are the most effective and accurate approaches and are widely used.
  - Biologic Approaches
  - Bean selection and genetic advancement
  - Frost resistance alteration in plant genetics
  - Chemical treatment of seeds
  - Flowering dates and planting dates (annual cultivation) (Perennial fruit crop)
  - Growth regulators and other chemicals A. B. Environmental approaches
  - Platform Reach
  - The variations in landscape and microclimate
  - Regulation of the diet of soil
  - Crop and Mulch Defense

Successful Protection

Constructive protection must be used right before and after the frost. Unable to avoid advective freezing, it is useful only in cases of radiative freeze. Good forecasts of the minimum orchard temperatures and the wind are very useful for the safety of effective freezing. Active safety depends largely on the elimination of heat leakage from the soil and plant surfaces.

Badness Season

A multitude of symptoms and plant damage is present, especially in cold weather conditions, leading to damage to branches, roots, buds, leaves and barks, wind scald and sunscald, bark cracks, drops and rolling leaves, and branch breakage. The root system, the lower trunk and the area around the crown are the most extreme. Damage to the region produces a moving layer fracturing that affects the flow of nutrients and can increase the incidence of redwood or insect injury. In the spring season, the diseased tree would not be able to absorb any more nutrients from the infected areas. But, after the energy is consumed, the tree will also be able to consume more nutrients. Winter damage, however, can cause tree destruction by insect pest invasion and disease in most plants if it continues to occur in the forest.

Winter destruction

Winter destruction typically occurs when temperatures fall below the critical level of each species. Wood is typically more cold-tolerant than flowering buds. Tree trunks and branch crotches are the slowest to harden, the most resistant to frosts. In late autumn and early winter, when the weather is cool, winter damage is incurred if the temperatures are not supposed to warm in a brief amount of time. Winter damage results in immediate tree death, lack of yield and reduced tree life, leaving the tree more susceptible to pests and diseases. The effect of the winter wound depends on the severity of the different factors, including exposure to cold weather, plant hardiness, plant ability, soil moisture, irrigation, plant location, variety of cultivars and plant sections, etc. Roots, planks, branches, buds, etc. Winter injuries are more serious at lower temperatures and vary from field to field. The tree would be more likely as the pressure rises with moisture deficits and hot wintertime, leading to plant failure or poor fruit set. The following signs of winter damage are present in the forest:

- The weaker performance of old trees
  - Beginning of buds of flowers and plants
  - Damage to the tips of the trees
  - Some vines in the soil destroy grape buds, cane and trunk damage.
  - Out of the tender cultivars Jonagold and Mutsu, the late leaf.
  - Cold patches of dying roots of young apple trees
  - Day-neutral low-growing strawberries and black crowns
  - Say raspberries and blueberries to kill
  - Tissue or part of contaminated plant dehydration
  - Distribution of branch and trunk bark and exudation of sapphire
  - Wilting, soaking of water and browning of wood

- Dead leaves lying on trees and drops of leaves are also present in harsh weather conditions.
- Root, branch and trunk decolouration

Winter injury types

Winter injury can take many forms in fruit trees as described the following: Blackheart Winter injury is very common in winter, where gumming leads to cell mortality and the pit is often killed along with darkening of the heartwood. Roots and branches can become more brittle later, but the trunk may be quickly replaced in healthy trees. It's all about the lichens, the young trees, and the black spirit. It is mostly used for the cultivation of strawberries, fish, pears, plums and cherry berries.

- Damage to the Ecology

  Hot temperatures, followed by cold temperatures, increase the degree of damage to the cambium, which leads to the hardening of the modicum. As a result, the plant has been contaminated with secondary cancer and fungal infections. Rubber can also occur occasionally, but not always. Cambium damage is primarily found in stone fruit such as peaches, plums and apricots.

- Collar/Crown Fracture

  That causes a lack of bark near the field in winter which can delay damage to the collar due to hardening. It is most common in Northern Spy and Graven Stein cultivars.

- An incident with the groin

  In winter, the vertical divisions are more likely to be influenced by shallow cross-windows. Such injury can occur in the upper and lower branches. The angle of the groin and the lateral split should be higher to avoid crotch injury.

Drought damage

Drought damage leads to the browning, yellowing and infringing discolouration of evergreens during the winter or early spring. Dryness happens as the winds strip moisture from the leaf in the winter, and the moisture from the soil is still frozen, resulting in a lack of water. Broad flooring plants are more likely to desiccate than thin flooring plants. The exposure to drawing was due in part to the curling of the leaf and the browning of the blade margin. A chemical anti-transparent foliar spray (to prevent moisture loss) should be avoided to stop freezing soil moisture and the painting of leaves and wood chips at the base of the plant. Winter sunscald is also known as a southwestern injury because, later in the day when the temperature is high, the sun is the highest on its side, with trunks and tree branches facing south and southwest warm. The tree bark warms in sunny days, and the cells that had previously fallen asleep become active due to the temperature of the plant. In the winter months, damage happens as active cells are destroyed within the trunk and branches of vulnerable trees. Refrozen cells lose some of their coolness during overnight temperatures that can lead to bark cracking and discolouration as they dip below the freezing mark. It is present mostly in apples, which can be extracted by painting white latex. Trunk division typically occurs in late autumn/early winter with a rapid decrease in temperature. Cracking/splitting will close or rebound. It spreads around the pit and can be popular with apples and sweet cherries. This is also the "the beginning of the end." of the fruit of stone.

Death Shooting/the Dieback

If the plant is not properly acclimatized/hardened, freezing temperatures kill the tip of the plant, resulting in the death of the plant It has always taken place in young trees at the head of cutting. Nitrogen fertilization and late cutting should be done properly to solve the problem of dieback.

Heavy injury

The finding of plants grown mostly in deep clay and wet soils is a severe injury. Alternatively, soil freezing and thawing can cause flaky plants to arrive. Mulching applications before soil falls late in the fall are used to address extreme soil and heat fluctuations.

Death of Root

The top part of the tree is impacted instead of the bottom part of the tree.. The sudden drop in temperature after sunset causes the bark to cool, compress and crack quicker than the wood. The continuous heating and cooling process lead to the isolation of the trunk.

Damage to the lowest temperature

The period can kill or subzero the fruit plant and ruin buds, branches of non-vigorous trees and shrubs, if it lasts longer. Inadequate soil and tissue moisture plants are very susceptible to low-temperature disruptions. Defoliation and premature cutting can result in low-temperature injuries in the growing season. Low-temperature penetration is not reversible and cannot be removed until it happens. It is common in fruit cultivation, like peach, apricots, cherry and nectarine.

Snow and frost injury

Structural damage to fruit trees and shrubs results in significant quantities of snow and ice combining with the wind. When the branch breaks, not all broken branches can be replaced, and the ice deposition is quickly cut into rough timber so that the trees are particularly durable and the ice can be removed more permanently. If new buds are weakened by late spring freezing, most plants may bud early, except some evergreen trees, and frosty buds will usually be lost for that season. Late spring frosts produce tiny fresh leaves that appear to be packed with holes between cups, turns, loops, crunches or shoots. The scale of the leaves is extended by the infected leaves as such species are spaced out. These incidents do not impact plants permanently.

Figure 1.5: Frost Injury and Ice Damage

Source: https://www.missouribotanicalgarden.org/gardens-gardening/your-garden/help-for-the-home-gardener/advice-tips-resources/pests-and-problems/environmental/winter-injury/frost-injury-and-ice-damage.aspx

Steps taken to mitigate injury

The following steps must be taken to mitigate or prevent a lack of mitigation during the injury season:

- Less cold-friendly places are selected.
- Heckles are also diluted.
- The option of cultivars of hardy and rootstock in local soil and climatic conditions.
- Limit nitrogen applied to soil during the early season.
- To conserve soil moisture, add organic loose mulch to the root zone.
- Take-cff, fertilization or late-season irrigation, which can encourage new growth, should be avoided.
- Take cut the wood that is dead and seriously injured.
- Get shade or windbreak in the winter months.
- The white latex tree trunks are painted.
- The use of wind gear.
- Right and timely spring, summer and fall irrigation.
- Taking as much fruit on the wounded trees as possible, so that the tree is regenerated instead.
- Usage of ice frames, greenhouses, fireplaces, smudge pots, aerial watering and light bulbs.

Winter damage associated with different fruits

Apple fruit

Winter injuries are the greatest barrier to fruit cultivation. Hardy MN apple trees can tolerate extreme cold conditions in the middle of the winter. The buds continue to expand during sleep breaks, which ensures that they lose the capacity to survive harsh winter weather. Swollen buds survive freezing temperatures without suffering damage in the early stages but damaging those buds at a temperature of about 20 °C. Winter damage could lead to dead flowers, leaves, shoots, branches immediately after the transition of dormancy. In young plants, low-temperature injuries to dwarf and semi-dwarf roots are most often attributed to late rainfall and high temperatures late in the season, and early dormancy due to lack of moisture and available soil water. Apple roots tend to get less acclimatized at low temperatures and the root is killed below-8 °C. Cold weather damage to the root of apple trees can occur in northern Canada and the United States. The next season, with the leaves leaving the trees, shows symptoms of root failure and new growth wilting and dying in serious circumstances. Cold hardness or subfreezing temperature tolerance of rootstock plantings in U. S. Orchards, though B-9 rootstocks had higher cold hardness.

Figure 1.6: apples

Source: https://www.healthline.com/nutrition/10-health-benefits-of-apples

Various attempts have been made to calculate the relation between winter injury and environmental (temperature) variables, and criteria have been set for calculating the susceptibility of apple trees to winter injury based on the minimum monthly air temperature and the defined rate of change in air temperature. 'Winterkill Day' is a snow day of fewer than four c.m. with a minimum temperature of-150 °C (moderate) to-200 °C (severe). Buszard observed that early cultivars suffered less from winter injury because they started to sleep sooner in early tree harvest than in late cultivars, while Quammeet found that apple trees were more likely to freeze during their acclimatization time.

Banana

In regions of freezing temperatures, edible banana plants are not stable, whereas ornamental banana plants are much more cold-tolerant. The degree to which bananas are affected depends on the minimum temperature and the level of exposure to freezing temperatures. Bananas are extremely susceptible to winter injuries and produce only edible fruit with cold frosts. During the moulding period, if any light frost is present, the maturation process may be stopped when hard frost slows down the mixing process continuously by grinding the fruit into black and mushy. The leaves show signs of burning when light frosts arise, but the infected plants will produce new leaves to replace the leaves.

Figure 1.7: A banana

Source: https://www.livescience.com/45005-banana-nutrition-facts.html

Infected leaves may be removed as fresh leaves grow. Sit back and wait for temperatures to get cooler, and new shoots will take place. The flower of the rhizome. If too many cold temperatures existed, it will kill in the winter. The roots below the field will later kill the whole top portion of the banana plant. It will recur in the spring if part of the roots stays. The Japanese fibre banana is a solid 20 F banana plant.

Blueberries

One of the limiting factors for the development and development in the production of blueberries is winter injury. Young plants are much more delicate than older mature plants. The branch tips of the blueberries are most likely to have winter injury. Moderate winter injury only leads to a few dead trees scattered and also lowers fruit production, as plants can die from serious winter injuries. Rough stems are more vulnerable than slow-growing stems to winter disturbances (one or two feet in height during the summer). Winter-sensitive branches must be cut from live tissue. For blackberries, nitrogen does not fertilize until mid-June. The branches will begin to develop in autumn if at the end of the summer the plants are given nitrogen Broad moisture with irrigation or rainfall during dormancy. Seasonal injuries can also be minimized in areas sheltered from winds by rising blueberries in winter. Winter disruptions, frost or a breathable mulching row can cause damage to young plants. Row coverings should be stretched over the berries in November and discarded within a few days after the snow melts. The straw fungus should be well handled and the mice should feed on the blueberry bark to lure them.

Citrus

The trees are evergreen in nature and never enter the dormant stage, except Trifoliate Orange. The freezing of these plants is vulnerable to cold damage due to large citrus leaves when conditions arise around the harvest. Compared to older trees, trees below the age of five are more vulnerable to frost damage. The site should be free from frost and cold winter to avoid cold harm caused by citrus fruits and prevention steps to fix the problem. If the minimum annual average temperature is between 25 and 30 F, most citrus varieties will be perfect. The ornamental orange trifoliate, which develops unspeakable fruit and performs well at 5 F, is the most resistant of citrus fruit, white grapefruit, tangerine and sweet orange are high at temperatures below 20 F, with no apparent loss. The coldest resistant plants, i.e. orange and lemon, show harm as temperatures fall to 25F. To avoid winter injuries, the citrus tree plantation in the south, which destroys the frost layer, must be completely sunlit. In winter and autumn, it should be prohibited from cutting and fertilizing because it stimulates new development of flushes, which is more likely to cause cold harm. Small trees with burlap bags or blankets can also protect from short-term freeze covering. Despite the thaw, the freeze weakened the leaves of the citrus tree and became rigid and brittle. The plant protects the leaves if the injury is not serious, but the bleeding as such continues. The seriously wounded citrus leaves slip out of the tree and dry for a few weeks. There may be a dim hue, rotting and dead bark patch, as the signs indicate on the twigs and branches. If citrus wood is badly injured, the leaves will hang on the tree for a long time, while the leaves will quickly fall out of the tree if the wood is damaged (K. K. Jindal & R. C. Sharma (prof.), 2004).

Grapes

Winter losses include direct losses in grape processing and the production of value-added products, that is, wine. In Finger Lakes, for example, winters resulted in a direct plant collapse of $5.7 million and a value-added estimate of $41.5 million in January 2004 for damages in missing wine sales, while total wine industry losses of $63.6 million were recorded from one freeze in the first quarter of 2004. In the phase of inactive grapevine, the fabrics that survive intense freezing display a light green or creamy green hue. During frost disruption, the membranes around each cell and the subcellular organelles can be disrupted. The winter condition occurs and extends from the interior of the phloem to the wires and trunks of the phloem tissue. It is more susceptible to winter loss and there is no indication that the wine is affected in the winter. This is attributed to the severe effects of the winter damage on the grapes.

Blind nodes mean that the secondary and tertiary buds have been damaged by a complete breakdown of the bud break. Injuries to the canes, cords and trunks of the vascular system often result in significant winter injuries to the buds. This means that the bulk of the shoots are unable to collect enough water and nutrients to prevent the water from sweating. The plant canopy will wilt abruptly and die at the end of the spring or summer unless the vascular system helps the vine to spread. It is difficult to recover from the winter injury the tragic damage to the grapevine trunks, though thorough planning and tapping could be required to cope with partially killed grapevine, localized cane and cord wounds. After severe winter disruption to the overgrown fields of the vine, the clandestine can still hit dormant and drive like fast-growing shootings at the base of canes or old woody trees. Even, viable base buds, which have been inactive for a long time, can emerge as very vigorous suckers if the entire canopy is destroyed by a winter wound (K. K. Jindal & R. C. Sharma (prof.), 2004).

Mango

Mango trees are the most cold-prone fruit plants raised widely in subtropical and tropical climates. The leaves and twigs of the plant suffer serious harm if the temperature drops below 30 F. In contrast to low temperatures, the high temperature is less toxic to the vine. Frozen damage happens when ice is formed in plant tissue and kills plant cells. Cooling threats get more severe when combined with cold winds.

Figure 1.8: Mango

Source: https://www.freshplaza.com/article/9183521/sri-lanka-new-mango-variety-discovered/

Olive

Defeat, bark split and branch dieback caused the most damage to olives. The most common signs are leaf chlorosis and olive knot infestation in frozen tissues. A lot of fruit buds have been killed by the frost, and a lot of fruit has been poorly or wrongly produced in the absence of leaves. The final economic impact is still unclear as a result of the freeze. Excessive irrigation can promote more vegetative growth, particularly after harvesting, which can cause more damage to trees. On the other hand, insufficient water will promote the drying of plants, which improves the hardening and decreases the growth of vegetation.

Figure 1.9: Olives.

Source: https://www.healthline.com/health/cold-pressed-olive-oil

Chapter 2

Classification of Fruits

CONTENTS

2.1 INTRODUCTION

In the cycle of the reproductive life of fruit, the ovarian of the tree is an integral auxiliary structure. Since fruits are extremely consistent, they play an important role in the classification of angiosperm, particularly when cultivated under different conditions. In general, the term fruit refers to the juicy nutrient structure used in everyday usage for example as apple. Fruits are commonly referred to as vegetables such as *Phaseolus vulgaris*, Eggplant (*Cucumis sativus*), Okra (*Cucurbita* sp.) and Squash (*Cucurbita* sp.) and Cucumbers (*Cucumis sativus*), and corn (*Zea mays*) and oats (*Avena sativa*) and other grains.

While most of us have an excellent idea of what fruit and vegetables we eat, it is difficult to explain precisely what carrots make one food and another fruit. It is simpler for the botanist to say that the fruit is the structure of the angiosperm with which the ovarian and accessory tissues are covered and the seed is formed.

Fruits are also important for seed dispersal. Vegetables are part of one of the plant's vegetative organs: roots, stems or leaves, fire systems. There are some complex 'vegetables': broccoli and cauliflowers are inflorescence buds, and the entire inflorescence is artichoke. Since flowers produce fruit and seeds from the reproductive organ of the plant, some inflorescences can be much more similar than fruits. An example of an inflorescence fruit that we call is pineapple.

The development of seed and fruit is the outcome of the fertilization process. Ovarian tissue is exposed to a series of dynamic changes in the growth of the fruit as the seeds fall out of the ovules. Several fruits are "fleshy" and contain sugar, which attracts animals and the animal digestive tract passes around and spreads seeds to new locations.

The wall thickens and becomes more or less distinct in three layers as the ovary expands into a fruit. The pericarp that covers the seeds or seeds that grow together form three layers.

There are three layers of fruit available:

- Exocarp, primarily the final layer of epidermis,
- Mesocarp or central disc, different in size
- Endocarp, showing a significant difference between species

Exocarp (a.k.a., epicarp) refers to the outer surface of the pericarp, the shell, the rind or the peel of the fruit that protects the mesocarp. The middle layer of the fruit, usually the fleshy plump part, is called Mesocarp (it comprises funicules in grapes, threads that attach the seed to the pedicel of the grape). Mesocarp contains alb in some fruits, including citrus fruits (In Latin, albus means white.)

Figure 2.1: Fruits.

Source: https://www.diabetes.co.uk/food/fruit.html

The Nature of the Ovary Determines the Structure of the Fruit

While the family lines differ greatly, the fruits have basic developmental and anatomical characteristics. For example, all fruit grow and differentiate into three layers of the ovarian wall which is induced by fertilization. The fruit wall (ovarian wall) is more or less distinctive layers.

Endocarp is the name of the inner layer that can be formed by layers that cover and secure the seed. Compare the heart, the rock, the peach, the rose, the plum, the olive, the hotbed, the walnut or the orange. It's distinct in shape. -Any curve, shape or shape is strange. Endocarp, mesocarp, and mesocarp, respectively. Fruit, like pedicel, calyx, dried leaves, stigma and corolla, also have floral structure. Manufactures that are part of a fruit rather than an ovarian plate are called accessories. Pineapple, oranges and strawberries (*Fragaria* sp.) can be used as pineapple accessories (*Ammas* sp.). A dense achen pulp receptor is the food component of the strawberry (dry, single-seeded fruits).

2.2 CLASSIFYING FRUITS

Fruits can be either simple or compound

There are three main types of fruit available. Simple fruits are produced from a single ovary.

More than one fruit is produced from fruit compounds. There are two types: composite fruit and multiple fruit. A single floral with several distinct ovaries is made of aggregate berries, all of which are attached to one container (for example, strawberry).

A variety of fruits have swollen ovaries with different flowers rooted more or less in a single mass, such as pineapples. The receptacle is extended by a vast variety of fruits, such as figs (*Ficus* sp).

The following conditions are considered when classifying compound fruit groups:

- The unit that grows the fruit.
- Total of mature fruit-forming ovaries.
- The number of human ovarian carpets.
- The essence of the maturity of the pericarp (whether the fruit wall is dry or fleshy).
- If the pericarp is split at maturity (dehisces) or not.
- The way the pericarp splits, whether it dehiscences.
- The location of the accessory tissues in the production of ripe fruit, if any.

2.2.1 Simple Fruits.

Simple fruits emerge from the mature ovaries in a single flora. Simple fruits come about in several ways. They can be dry or fleshy in their pericarp (fruit wall). The pericarp may or may not be desiccated until it is dried.

2.2.3 Fleshy Fruits, pericarp fleshy at maturity

For food, the fruits in this category are common.

They have a fleshy wall of fruit (pericarp). The fleshy portion is generally appealing to animals, who eat the fruit and in exchange take away the seeds. The seeds present in these fruits appear to have a stiff seed coat that is not broken down when the seed moves through the animal and is collected in its feces.

Examples of drupes are plum, almond, peach, and apricot (all are Prunus sp.), within the rose family (Rosaceae). There is also a drupe in the olive (Oleo sp.) (Family Oleaceae) fruit. The drupe, originating from a single carpel, is typically one-seeded. It has a rough endocarp composed of thick walled sclereids, forming the skin with a thin exocarp. The edible fleshy part is the mesocarp. The seed, with a thin seed coat, plus the stony inner coating (endocarp) of the ovary wall, is the pit of a cherry. In almond fruits, when the fruit is young, the mesocarp is fleshy like a typical drupe. However, as it grows, the Mesocarp becomes hard and dry and the hull emerges. The almond shell is an endocarp. This is a case of which the edible part of a drupe is the seed, not the outer part.

A berry, produced from a compound ovary, is a fleshy type of fruit. Typically, several seeds are lodged in the pericarpal flesh, but it can be difficult to see the demarcation line. To us, it comes as a surprise that tomatoes, lemons (Citrus Limon), and cucumbers (*Cucurbita* sp.) are berries, but unlike their common names, strawberries (*Fragaria* sp.) and blackberries (*Rubus* sp.) are not. There are strong examples of non-matching botanical names and generic names.

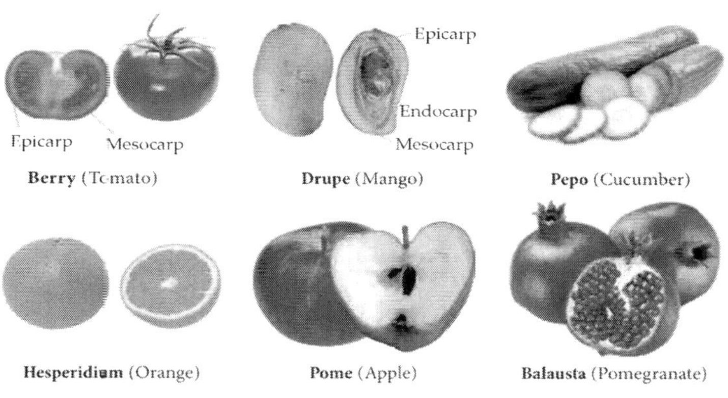

Figure 2.2: Simple Fruits and its Types: Fleshy and Dry Fruit

Source: http://www.brainkart.com/article/Simple-Fruits-and-its-Types--Fleshy-and-Dry-Fruit_32954/

A type of berry called hesperidium is lemons, oranges, limes, and grapefruits (all *Citrus* sps.) The thick, leathery rind (peel) with various oil cavities is exocarp and mesocarp; several wedge-shaped locules are made of the thick, juicy pulp segments (endocarp). In juice sacs or vesicles that are an outgrowth of the endocarp walls, the juice types. Many living cells filled with juice are comprised of each mature juice sac. Watermelon (*Citrullus vulgaris*), cucumber and squash fruits—all members of the Cucurbitaceae family—are a type of berry called a pepo. The fruit's outer wall (rind) consists of receptacle tissue that covers the exocarp and is fused with it. The fruit's flesh is mainly mesocarp and endocarp.

Pomes are examples of apples (*Malus* sp.) and pears (*Pyrus* sp.), all of the Rosaceae family. This fruit is produced from an inferior ovary flower. The flesh is inflated hypanthium (a fleshy floral tube), and the heart is from the ovary.

2.2.3 Dry Fruits, pericarp dry at maturity

Achen is a dried fruit of only one grain. Sunflower achenes (*Helianthus annuus*) are generally referred to as seeds, but when the pericarp has opened, the seed is carefully seen inside. Pericarp may be separated from the seed layer, which is a thin coating to protect the embryo of the sunflower. Caryopsis is the fruit of the rice (*Oryza sativa*) and wheat (Poaceae) grass families (*Triticum aestivum*). The grain is a young, single-grain fruit, unruly. As the seed coat and pericarp are closely interwoven in the embryo, it differs from the achene. Samara can be a single seed of simple fruit, such as Elma (*Ulmus* sp.) or two seeds, such as maple (*Acer* sp.). These fruits are distinguished by the presence of the ovarian wall, which forms a wing-like structure that helps disperse seeds. The fruit of the Apiaceae family is the schizocarp and the celery (*Apium graveolens*). The schizocarp consists of two rugs which, if ripe, are divided into two seeds and are in the middle of the rug. The word nozzle for many hard-boiled fruits and nuts is widely used. The typical nozzle is a single, undisclosed, dry fruit and a rough or stony pericarp (shell). Castanea sp (*Castanea* Sp) is an example of a chestnut tree (*Juglans* sp.). Granite is partially filled by a cup of hardened oak (*Quercus* sp.) fruit. The bracts, the perianths and the outer layer of the pericarp are the outer shell of the walnut that is discarded during fermentation. The raw coat is the remainder of the pericarp. Remember that the unshelled almonds are not seeds, but fleshy fruits known as drupes, which have been removed from hulls such as exocarp and mesocarp. *Bertholleda excelsa* (Brazil nuts) are beans and not nuts, and *Arachis hypogaea* is a pod.

Dehiscent fruits

These are those which, when fully mature, dehisce or break open

- Follicle, consisting of one carpel and separated by a single suture,
- Legume, consisting of a single carpel and separated into two sutures,
- Capsule, consisting of multiple carpels and opening in one of four directions at maturity:

Indehiscent fruits

There are those that do not crack at maturity.

- Achene or akene, a single seed fruit with only one seed point attached to the fruit.
- Caryopsis or grain, a single seed fruit in which the seed is closely bound to the fruit in all possible times.
- Samara, a single or two-grained fruit with an outgrowth-like pericarp bearing a wing. Achene has shifted
- Schizocarp, composed of two carpels, which are divided into two single-seeded halves around the midline at puberty, one of which is indehiscent;
- Loment, having several seeds, dividing at maturity into single seed segments.
- Nut, a firm, single-seeded fruit, typically produced by a compound ovary, with a hard pericarp inside.

2.2.4 Aggregate Fruits.

A collection of mature ovaries, made up of an aggregated fruit and formed by a single flowers over the superficial of a single receptor. It is understood that fruitlets are human ovaries.

The fruit of one flowers is an extra fruit, but there are multiple ovaries in the flora and the ovary is fused together as it grows. A plain fruit, not a whole fruit, is the fruit from a single flowers having just one ovary. Not all the flowers become aggregate fruits with several ovaries; many ovaries are not closely related in order to produce greater fruits. The fruits can also be attached to the accessory. Sections of the flora that are not ova, become juicy and become part of the fruit in accessory fruits.

Figure 2.3: Aggregate Fruit

Source: https://study.com/academy/lesson/aggregate-fruit-definition-examples.html

In general, the fruits contain:

- Blackberry
- Strawberry is a fruit composite and an accessory. The ovaries of the strawberry are the tiny dark spots. From the flower base comes a juicy portion of the strawberry.

2.2.5 Multiple Fruits.

The adult ovaries are produced from multiple fruits and are made up of several or more flowers, more or less mixed in mass. Numerous fruits are accessory fruits almost always.

Multiple fruits, which consist of a cluster of fruiting flowers, are fruit bodies called collective fruit. Every flora in the inflorescence produces the fruit, but it becomes a single mass in which all flowers bear true fruit. At the flowering, the mass is considered to be a fruitlessness. The mangoes, mulberry, Osage orange, breadfruit are examples of these.

In languages other than English, numerous and absolute significances of fruit are inverted such that several fruits combine into one flower several pistils. The flower of certain plants, such as this noni, is frequently developed, and examples can be seen in a single branch of flowering, fruit growth and fruit maturity (FAO, 2020).

Flowering and fruit growth in the Noni or Indian mulberry can be seen in a single branch (*Morinda citrifolia*). Next, there is an inflorescence of white flowers called a crown. After fertilization, each flower transforms into a drupe, and when the drupes ripen up they transform into the known fruit known as the syncarp. A variety of dried fruits are also available.

Other various fruit representations:

- Flat tree in one fruit shell, several multi-floral achenes.
- Mulberry, one fruit consisting of numerous blossoms
- One fruit, jackfruit, is several flowers,

2.2.6 Accessory Fruits.

Accessory fruit is classified as fruit that comes from ovary-covering tissues.

The accessory fruit typically matures in flowers with lower ovaries, while the seeds contain the receptacle or hypanthium. Simple, or individual fruits may be used as an accessory. Accessory fruit, which is also called the pseudocarp, is a misleading fruit. Sections other than the ovaries and their walls are essential for producing a portion of these fruits' fleshy fruit. The actual fruit is red, the accessory is white.

The real fruit, Dry Achene; the accessory, when ripe, is red and juicy. The basic fruits of the lower ovary are several of the complementary fruits. For e.g., grapes, strawberries, quinces, pineapples, melons, and other accessories.

Strawberries

As mentioned earlier, strawberry is an accessory fruit; in other words, the pulp is not a part of the ovaries (which are the "seeds" actually achene). However, it may be logical that the seeds are the true fruit of the plant and that the strawberry flesh is a vegetable.

As it grows, it is greenish-white and in most plants, until it matures, turns red. Rosettes are known for the development of these plants. The majority of species offer their extremities slender long runners that produce a new bud. The leaves typically have three leaflets, but there may be five or one leaflet number.

Apple

It's a leafy flower, The Apple Tree. The spring season is suitable for the flora and leaves of apple bulbs. The flowers are white, with five petals, the first tinted rose. Simple oval on a peccliola of 2-5 cm with an acute tip, a clamped margin, 5-12 cm long and 3-6 cm wide. The leaves are alternately arranged. The fruit season is typically in the fall. "An apple a day keeps the doctor away" An apple keeps the doctor away every day. We think so for a variety of reasons! Apples are recommended to decrease the risk of cancer of the intestines, breast and lung.

Watermelons

Watermelon, a vine and fruit of a wine-like herb derived from Southern Africa (brancher and trailer). This flowering plant has an accessory like those simply referred to by botanists as a berry. Watermelon fruit is a smooth, ripe, juicy, normally red inner flesh, loosely assumed to be a kind of melon (although not in the genus Curcumas).

The gene descriptor of *Catullus vulgaris* is also synonymous with this herb (*vulgaris* meaning "common"). A host of alternative ways of drinking fresh sweet watermelon can be added to hot summer drinks and smoothies. The easiest way to cut an watermelon is to cut it in half or quarter and then cut the resulting circular plaques. This technique is commonly used in a comfortable atmosphere in which people do not worry about juices going anywhere.

Pineapples

The pineapple is a fruit from the Asian tropics with its fragile and fresh scent that is totally addictive. The fruit top is like the royal crown and as some might have argued, the legendary Phoenix eagle's wings, and the flesh is like the Asian pear's hide. The word pineapple in Taiwan sounds like a good health and successful blessing for the future.

Decorating your house or workplace with symbols depicting a fortunate pineapple is a common ritual that assures that all of your needs can be saved and all of your goals achieved. Conventional culture has suggested that the pineapple is the perfect gift for a house warming party and the initiation of a new venture, or that it wishes a chosen candidate luck in election boxes.

Pineapple has long been introduced into Taiwan and in the 19th century was a popular fruit on the island. After the end of the Japanese rule in Taiwan, it became a major producer within a decade of vigorous melting pots of cross-bred pineapple, promoting pineapple products in the region around Chiayi County.

Since pineapple has high temperatures and outstanding drought tolerance characteristics, it is suited for pineapple propagations and areas where temperatures differ during the year and throughout the day. The spring is best suited to the pineapple's natural habitat and weather requirements. A vast number of pineapples, including atemoya ice pineapple, winter sweet pineapple, ice cream pineapple, scented apple pineapple, perfume pineapple, etc. have been produced. Pineapples are not long-term processed in luxury and are normally harvested early for exports to reduce quality losses during shipping. The bulk of pineapples are shipped from Taiwan to Japan, Hong Kong, Singapore and Canada.

Berry

A berry is a plain fruit made of seeds and pulp by a single ovary. In non-technical uses, berry denotes any small fruit that can be eaten whole and which lacks irrational seeds.

Bramble fruit is one of the most popular Rabus compound pseudoberries (blackberries). Many varieties of berries are known as true berries and fake berries.

2.3 BOTANICAL DEFINITIONS

The term botanical fruit typically refers to the mature ovary of a plant, with seeds, cover and any tissue closely related to it, regardless of whether it is edible in a specified manner. The term fruit in reference to food refers to the edible part of a plant consisting of seeds and tissues. Seeds that last on the plant (such as blueberries, cantaloupe, poach, pumpkins, tomatoes) or dry seeds where the ripe wall of the ovary is composed of parchment, leathery or woody beans, pulses and nuts. In the broad sense, in its botanical meaning, the term vegetable refers to any weed, edible or inedible, including trees, bushes, vines or vascular plants. In the usage of vegetables for non-chemical seasoning items, the word implies plant material used for food. Vegetables are the edible part(s) of the plantation, such as the herb(sprinkler), the root (potato), the pulp (potato), the leaves (spinach), the flower(s), the fruits (apples, cucumbers, pumpkins, strawberries, tomatoes) and the seeds, according to one species, the edible part(s) of the plant (beans, peas) (beans, peas). The fruit that falls under the latter classification does not include vegetables.

We have to choose one in four systems used to divide living organisms between the kingdoms of fungi and algae (foods used as common vegetables) if mushrooms and algae (foods used as common vegetables) are deemed to be part of the plant kingdom. In the kingdom of plants, fungi in the kingdom of animals, and animals accommodate algae, all of whom are the food source of mushrooms in the kingdom of plants. Three medium-sized DNA (deoxyribonucleic acid) bars were divided into three different kingdoms, including the Kingdom of Protista (FAO, 2020).

2.4 BASIC DEFINITIONS OF FRUITS.

Fruit, grain, seed, and nut types are most important for food groups, which are the most important plant foods. (Smaller groups include the classes such as coffee, tea and chocolate). Herbs or herbal medications shall be used. Plant foods and the types of foods made out of these foods are used to communicate about all of the different foods food groups. There are these groupings of cooking devices that are used in homes, in schools where nutritional workers convey the cooking skills of their customers, on the sidewalks, and in restaurants where their customers purchase and eat cooked food. The list of cooking foods available in the EU consists of only 24 items, including meats (including poultry), fish, eggs, milk, some quantities of legumes (beans and lentils), wheat products, potatoes, spices and oils, and foods and beverages that contain such items.

If you include cereals in this list, you cannot say you're referring to boiled rice, pasta and potatoes, because these are only grains. If you are referring to fruit and vegetables, for example, you can say, 'Cooking fruit and vegetables is a broad term that refers to fruits and vegetables that are used as 'grains' and meats that are used as 'pasta'. 'Grains' can also include cereals, seeds, nuts and peanuts.' Fruit and vegetables are presented in this research as having specific, specific qualities as a result of them being small in size and lacking in the fat content to prevent the perception of shelf-life in comparison to most fruits. The word fruit is a food type, made up of a sweet or sour taste, and another part that does not taste good but is still nutritious. Under U.S. law, culinary fruits represent a subset of botanical products that remain after the removal of other plant materials (which are wheat, rye, oats, and barley). Everyday life is made up of a mixture of morning drinks, side dishes (such as orange juice or mango), coffee, pizza, lunch, and dinner foods. Fresh fruits and vegetables are also used as food and ingredients in salad and side dishes. Veganism, a dietary style of vegetarianism practiced by some groups of vegetarians that focuses on plant-based foods, is defined as a philosophy and way of living which seeks to exclude all forms of exploitation of, and cruelty to, the animal kingdom. The foods which may help to support are the foodsthat have come from the plant family (such as flour, wild rice or millet, oats, barley, avocados, peanuts, legumes, olive oil, or chocolate) (cereal grains used as vegetables). The food item "vegetables" wouldn't be good for people who aren't alright with other grains, nuts, peanuts (a form of pulse), and culinary fruit. The distinction between culinary vegetables and botanical fruits is largely a matter of the culture of foods, and there is a bitter taste to most of them. Say a tomato or a green pepper, for example, is usually sweet (due to the higher concentration of sugar) or bitter. Sweet is better appreciated with the brain, at least among most people because the brain likes sweet things (due to a higher acid content) (Valpuesta, 2002).

Traditional uses are the basis for so called culinary variations, which are used in the use of plant parts, such as fruit and vegetables, and tend not to be precise, different and/or inconsistent between crops (and those which are known as fruits and vegetables). In cookbooks and in public health organization nutritional guidelines that is published. This is not published in cookery books and in food manuals for customers. It is published in specialty cookery books and in food manuals for consumers. In order to improve or maintain health, a dietary educator would need to follow a diet plan that would allow them to find the optimal amount and type of food that would fulfill dietary requirements, prevent diseases and reduce the risk of chronic diseases.

Recent research by the food guides used in Australia, China, Canada, Germany, Mexico, the Philippines, Portugal, Puerto Rico, Sweden, the United Kingdom and the United Kingdom have shown that, given the cultural variation in dietary systems, food types, such as cereals, vegetables, fruit, and meat and meat substitutes, and milk products, and sweets, are largely similar. Food groups as a single fruit and vegetable category, six food guides (Korean, Canada, China, Mexico, Portugal and the United Kingdom) are included, but separate categories are included in the other guides. There are designated areas that are meant to separate both fruit and vegetables from nuts and seeds. The guide for starchy vegetables varies. For example, the tuber vegetables are more often recommended for the mature subjects. There are sixnations involved in the potato division in the vegetable community (Australia, Canada, China, Philippines, Puerto Rico, USA). Having the same taste as cereal, both root vegetables and tuber vegetables, such as potatoes and carrots, are listed as cereal in the German, Korean, Mexican, Portuguese and United Kingdom vegetable classes. Potatoes are often classified into the category of grain products because they: are starchy, are not costly to buy, and are often consumed widely and easily. In addition to potatoes and other root vegetables, the Swedish Food Guide recommends root vegetables as the foundation for a cheap and accessible diet (made available to you on a daily basis) to be complemented by other vegetables in a variety of ways each day and seasonally. Seven food guides put their protein related foods in the "meat" category; Australia, Germany, and Sweden do likewise for the carbohydrate, vitamin and nutritionally based foods in the "vegetables" chapter.

The American Food Guide gives schedules for the immature pulses, fish, meat, and poultry from the vegetables and the mature pulses. The Chinese Guide puts pulses point to the milk and milk products division in this case (mainly soya and soy milk). Vegetarian and vegan diet books contain materials on the function of proteins, have an emphasis on iron and other meat nutrients, and features numerous recipes that spotlight dark green leafy vegetables and dried fruit as the main ingredients. Food manuals for candies or sweets, fruit and vegetable food such as preserves, jams, and jellies, fruits, and vegetable sausages used as candies and sweet cucumber pickles. In addition to mixed dishes and sweets, no foods containing fruit or vegetables shall be held in the Food Guides for condiments, for items extracted from fruit or vegetables, or for herbs and spices. An illustration of this is that botanical descriptions of fruit and vegetables are more reliable than culinary explanations. This information and its effects cannot be understood by culinary experts, but instead must be learned (and respected) by your scientists and tasted by your food experts. There are a few things that can be affected by the following botanicals and culinary issues: berries, spices, seafood, fruit and vegetable mixtures, desserts and fruit and vegetable foods (Valpuesta, 2002).

Peppers and onions are vegetables that I sometimes bring with me when we go on hikes. Whether or not they are taken from plants is not always entirely clear, and depends on the perspective of the classification scheme used to classify species.

In agriculture, potatoes and other tuberous and starchy roots (e.g. taro) are typically placed alongside other crops and grain categories classifications of many crops and are considered a specific category separate from other grains along with other tuberous and starchy roots.

The pulses plant, apples, and chai (cotton) can be used as substitute meat in some crops, such as sweet potatoes, rather than fruit (or vegetables). Instead of being categorized like vegetables, products that contain soya bean products such as tofu meat alternatives are instead categorized as high protein products (meat, fish, poultry, omelets, nuts and seeds). If you are lactose intolerant, you can choose soy milk like milk, or easily the calcium from it.

The peanuts that grow on trees are a wheel that travels through different cultural senses (e.g., snack food, part of a main dish, boiled side-dish, peanut butter, or peanut sauce). Peanuts are often called nuts, and are known as an excellent source of protein. New and sweet maize and hominy grain is commonly used as an ingredient in many traditional dishes (i.e., side dishes with a dinner meal). To provide a less formal way of saying "corn," the more common way is "maize", which is commonly used to refer to the cooked, sweet kernel-shaped cobs made of maize. Corn, tortillas and tortilla chips are made of corn, rice, and other staple grains. Chopped fruits and vegetables are not inherently unhealthy; they only have a high percentage of fat and need to be cooked a certain way to kill the harmful pathogens in them (avocados, coconut, olives). The food labels do not provide the detail that certain items can be classified to make them safer. Avocados and olives can be combined with joint, vegetables and fats. Herbs and or fruits and vegetables may be mixed in with coconut, such as a fruit or vegetable sliced, trimmed or dicing together (e.g., used in stews mixed with meats and other vegetables). In plant species, the herbs are produced in the cap of the plant.

The parts of the plant used for seasoning are the stem, leaves, and flowers (e.g. garlic and chili peppers). Herbal and organic foods do not appear in the Food U.N. Codex, mostly due to their use in very limited quantities, but may contribute towards important components of food. It is not important to analyze the total fruit and vegetable diets based on the fruits and vegetables that are used in a non-main dish meal. The food guide (food pyramid) is to be used to figure out what to eat or not eat, but they are not to include groups of mixed dishes or desserts since it may include fruit or vegetables (meat and vegetable casseroles, stews and baked fried food) (chocolate-covered raisins, fig bars, fruit pies, pumpkin pie, carrot cake). Just because you eat fruits and vegetables, which are typically healthy, that does not mean they will maintain their nutrients and can be fattening or sugary. On the other hand, however, many foods such as puddings, soups and sauces, jellies and fruit drinks are typically put together to make a different category of the term (candy or sugars) which have a specific instance for category (licorice or soft drinks). Farm, fruit and verdant subgroups the grouping of trees, fruit and vegetables subgroups on the basis of their food sources may be of value for epidemiological research. In order to make this "power fruit" better, it's important to concentrate more of the fruit's nutrients that are lacking like omega 3 is low, sugars are low, fat is low, sodium is low, cholesterol is low, and vitamins are low. The rest of the nutrients are rare and less suitable to the human body. Due to the large quantities of different food constituents in fruit, a specific fruit can be grouped in the category and not all of the fruit groups have been investigated, leading to a possibility of watermelon, tomatoes, and apples having more chemicals than bananas, cantaloupe, and lettuce. Certain nutrients, such as vitamins C and A, carotenoids, folacin, copper, zinc, magnesium, calcium, and flavonoids are primarily found in fruits and vegetables. Most of the components of e-cig liquid are made from spices, tastes and flavors. In order for us to determine the risk associated with food, the recorded literature regarding the ingredients contained within our meals would first have to be all collected and properly compiled.

Since food composition databases show evidence for the origins of thirteen carotenes (dark green leafy vegetables, deep yellow and orange fruits and vegetables, citrus fruits and cantaloupe) and vitamin C (citrus fruit, dark green leafy vegetables, cantaloupe) and folacinine, it is likely that other food substances may be found (dark green leafy vegetables, oranges, pulses). Data from the food composition and databases for other biological ingredients such as glucosinolates, indoles, and isothiocyanates; flavonols, flavonoides and other flavonoids, fruit juice flavonoids, phenolic acids, citrus fruits flavonoids, and carotenoids. The fruit and vegetables subgroups for the reporting and exploitation of the results of food sources as well as the subgroups for edible parts, color, packing and preparedness are given. Plant classifications address the major physiological characteristics of plants. Many botanical plant classifications are based on the morphology of plants and the result of the organization of the plant. The botanical classification goes from Kingdom, Section, Class, Subclass, Order, Families, Genus, Plant, Variety, Cultivar, and the most consequential of all, the strain. For example, 11 terms of the Gray Zucchini Summer Squash classification are: Vine, Spermatophyta, Angiospermae, Dicotyledonae, Cucurbitaceac, Cucurbitae, Pepe L., Melopepo Alef, Zucchini and Gray.

Relatives, families, and botanical groups also help when deciding the roots between different family plants, and also help in botanical technology to understand the plants between them, and also help in plant medication styles, insect elimination, and culturally driven sciences. They have botanical classifications (systematists) which are very useful for plant farmers. The study of botanical designations for foods probably does not yield the same results because foods that originate from the same botanical family might or might not contain the same levels of bioactive components. There are four plant groups (three of which contain foods that can be eaten) that contain organisms that are consumed by humans. These plants are part of the four main groups (the typical two-kingdom regime). They split the bulk of human food into separate parts (seed plants). All of the nutrients in the human body, including DNA, are found in the sperm sex cells, as well as in the two sperm cells that both arise from the same single origin within the same individual (Gymnosperm and Angiosperm). In the Angiosperms (Monocotyledonae and Dicotyledonae) and the Families (Families), there are around 93 orders and a total of 432 families (20 orders and 67 families for the Monocotyledonae and 73 orders and 365 families for the Dicotyledonae). Even if only a small number of plants are used as human food, thousands of different species of fruits and vegetables are eaten in the world, and by understanding the various cultivars and strains available for each fruit and vegetable, the number of fruits and vegetables available to eat increases to millions. Some families are found to have a very particular focus on the specific components of foods, vitamins, and minerals, while others have a very broad range, which can be expressed differently.

Peterson and Dwyer suggested that botanical classifications could be useful in assessing the possible existence of flavonoids in food where data on food structure are not available, but they recognized that Imprecise is likely to be a quantitative approximation. Figure 3 shows the different food elements: fruit and vegetables, and lists their botanical families with the corresponding food that includes them. The Asparagaceae family is one of only a small number of families (or groups) that includes one type (or principal) of fruit and a distinct segment profile in the Olive family (or Asparagaceae) (Oleaceae). Foods containing some or all of the following might be candidates for testing: Carrots, dewberry, Foxglove, ivy, Lark, Lobelia, peony, strawberry, tulip, violet, cardamom, dill, fenugreek, garlic, horse-radish, jalapeno, mint, onion, rosemary, sage, savory, tarragon, thyme, and turnip (Asteraceae).

The so-named family or class of foods originating from the gourd, Cucurbitoide, Solanaceae and Carotaceae families do not have the same food ingredient profiles. They are not consumed in the same manner as cigarettes. Cantaloupes, watermelons and some yellow fall squashes are part (vitamin C and vitamins A and B) of the family of squash and the family of gourds (13-carotene). The foods that have this effect are not the main precursors for most other members of the gourd family, such as boniatas, pork rinds and winter squash (NYW). In the night shade family, Chile peppers (3-carotene, capsaicin) can be discovered; sweet peppers (vitamin C, lycopene, if red); tomatoes (vitamin C, 13-carotene, lycopene); carrots (beta-carotene, Vitamin A and C); (not major sources of food components) (T. Radha & Lila Mathew, 2007).

The plant family that includes Avocados and folacin contains plants that contain medicinal and spice ingredients (cinnamon, sassafrass, sweet bay). Everything in the cart are (X-and [3-carotenes]; vegetable celery stalks; root celery and parsnips; and carrot-like herbal seeds (anise, caraway, coriander, dill, fennel, parsley). As a result, while certain fruits and vegetables found in some plant species have similar food components, not all plant foods can provide a sufficient amount of a certain food component. Since fruits and vegetables are selectively assigned to certain plants families, they share difficulties in classifying them as food, like for example, because some species from the same plant family are edible in different states of withered, ripe and unripe, or because some of them need certain ingredients to be eaten in order to become edible (e.g., beet roots and greens, turnip roots and greens, broccoli stems and flowers, chive bulbs and green tops).

It is the issue of botanical classification that does not apply to a single plant part, so it applies to the entire plant. Plants can be classified into climatic circumstances, where they thrive in water or soil (altitude, elevation, hilly, etc.) as well as the type of plants they produce in deserts (low humidity, temperature), tropical (extreme humidity, high temperature) or temperate (medium humidity and temperature) or tropical (high temperature and humidity). You will find lotus, taro, chestnut, convolvallis, and watercress intertwining varying aquatic plants throughout the ecosystem. Cacti and desert cucumbers include desert plants, such as cactus and yucca (buffalo gourd). Although humans may eat various tropical fruits, bananas, rattans and bread, there are other tropical plants that are edible as well (Goa beans). Plants that grow in the temperate zones can be separated, basing on their particular growing season. Because the medium temperatures 15-18° C are suitable in the production of crops, crispy broccoli, cod, carrots, chard, celery, pulp, garlic, kale, lettuce, mustard, onion, parsnip, pea, radish, spinach, turkey and white potatoes are suitable crops for medium monthly temperatures. Seasonal crops of great beauty (60-65F). Cold season vegetables are not good for the most ideal temperature of eighteen to thirty-six degrees Celsius (fifty-eighty-five to one hundred-ninety-nine degrees Fahrenheit). Kidney beans, cucumber, egg, onion and potato tend to survive in the cold, but they may have poor quality produce (T. Radha & Lila Mathew, 2007).

Beans, cucumber, cucumber, eggplant, onion, potato, tomato and watermelon are frost-intolerant. Various crops include seeds, fiber, and fruits or seeds that are useful in many for many recipes (mango, papaya, figs, avocado, etc.) Cold plants include ingredients for various dishes that are used and very much popular (chalk, brushwood, carrot, cloves, etc.) Warm plants provide a variety of nourishing qualities and certain benefits to the body (carrot, endive, etc.) (pumpkin, sweet potato, tomato, watermelon). In terms of the different tables and the synergism, they bring forth, grouping by crop season, habitat, or environment does not seem to be directly relevant to the composition of food ingredients and is not especially useful for nutritional epidemiology. The community of growers may limit their farming to specific conditions they can sustain and control, allowing them to have a more directed approach to their agricultural practices. The definition of malnutrition is less useful for people that don't have access to food and for those who don't have the storage to keep food for themselves either (freeze, canning). The plant's flowers emerge through a special point on the plant and the fruits are classified into three groups: empty calyx, a flower head with normal flowers (anthers), or a head that emerges bearing a normal or abnormal flower (calyx). The outermost layer of the fruit is usually like the exocarp, commonly called the peel. The mid-region of the fruit is known as the mesocarp. In the middle, there is the endocarp. The mature fruits of dry-berries (grains, boars, peas, and nuts) are classified as dry fruits, while those fruits with some fleshy partial mesocarp are classified as fleshy fruits at maturity (all others).

The flower can either form a single pistil or it can be developed into a fruit. Either way, a seed forms within the flower or the pistil turns into other parts. Pomegranates, drums, and berries are all shrub-like fruits. If you are a smoker you will have to buy your nicotine from a dealer, who will have the nicotine in one of these special flavors: apricots, raspberries, chocolate, chocolate, dates and avocados. Also, some cigarettes also have spearmint flavor. The flesh of the pomegranate grows from the reddish-brown sac around the fruit's ovary, and the endocarp that surrounds each seed is actually made of paper or leather, as in the case of mango, pear, and quince. Beers are made from the same organisms as olives, and normally have more than one seed. All three types of berry are berries, which are actually pepos and hesperidiums.

Like watermelon, the thin layer of skin on the mango fruit tender but hard, the shell of fresh peaches, cranberries, peaches skin, the peel of an aubergine feel free to rip, but the bitter inner flesh is good to eat. Cantaloupe, cucumber, cow, watermelon squash, and especially the very thick rinds of these are all some types of berries. When the fruit increases, the inner wall lining of the ovaries becomes stretched until they turn light, and then become deeply stained and swelled with juice. All members of the family, including plums, apricots, strawberries, cherries, and peaches, will have a taste (grapefruit, kumquat, lemon, lime, orange and tangerine) (T. Radha & Lila Mathew, 2007).

A single serving of a flower, the pistil, might be involved with berry production. What happens to the pistils that are on a plant as it ripens? The pistils become fleshy druplets. There are examples of these such strawberries, cherries, blueberries and raspberries. During the maturity process, when a cluster of clustered flowers are successfully plucked as a ripe fruit, several fruits are produced. The floral type, for example, is one of the few components of the pineapple, and each section is connected to the central center, which has a woody stem structure. The study of the division of fruit into classes on the basis of the flowers, although the classes are not specific to the consistency of the food component, is unlikely in epidemiological studies. The plant with the same name is a unique example of the fruit that come from the Rue family (citrus).

Fruit classification can be confusing for nutritionists and study participants, as the term botanical fruit is used for fruits not typically listed as wine, such as avocado, pineapple, cantaloupe, cucumber, date, grapefruit, kumquat, lemon, lime, orange, squash, mandarin and watermelon. Oh, no, oh no. No, no, no, no. T. The fruit T. Titi is found in the class of fruits called berries as well as fruit classes and fruits. Government agencies also use one of the grades of fruit and vegetables while monitoring the national food supply (availability) and information on food consumption. These plants are also grouped into the category of fruit and vegetables. According to the USDA Economic Research Service, the most recent statistics are available for each of the individual fruits and vegetables available exclusively within the USA. These statistics are used to report descriptive results on food consumption on a per capita basis, i.e. availability of food per person. For example, in 1999, the US Department of Agriculture grouping of food.

The classification of the types of fruits and vegetables shall be based on the varieties of the different fruits and vegetables and the variety of production. There are four types of fruits (citrus, melons, bananas, and tomatoes). There are also four kinds of fruit preparations (fresh, juiced, canned, dried and frozen). Vegetables and four types of vegetable include: dark green leafy vegetables, deep yellow or orange vegetables, starchy vegetables, and dry beans, lentils and peas (fresh, canned, frozen and dehydrated). The specific plants they eat have their own set of favorite foods, but both types are large enough to be able to recognize the most commonly consumed items. The edible fruit and vegetable part of the spices, stems, vines or trees used as food shall be divided into each segment. According to a study, certain plant tissue has high nutrient levels that are associated with this disease (e.g., leaves, stalks and Stems, roots and tubers, and pulses). This method of classifying a few food products is used in some food composition databases. These tissues are fully functional regardless of whether they can get the nutrients they need from various sources (Hancock, 2008).

Most vegetables, (for example celery and rhubarb) usually contain large amounts of dietary fiber needed to prevent cell plant structures from getting damaged. Leaves, especially dark green leaves, are usually the most metabolic, nutritionally abundant in plants and provide a good source of dietary fiber, folate, carotenoids, vitamin C and minerals such as iron, zinc, calcium and magnesium. Regularly, meat is fortified with nutrients, including high amounts of protein and B vitamins (mature beans and peas). The reason for the bulbs was known by Allicin (chives, garlic, onions, shallots). Larger root and tuber organisms produce a high volume of starch which is stored. A large root and tuber will store starch (potatoes, sweet potatoes, taro). Tubers and roots that have less energy compared to before (e.g. Jerusalem artichokes, parsnips, turnips) (e.g. carrots containing EURL- and f3-carotene). The variety of nutrients in fruit is higher than in vegetable fruits, pineapples, strawberries, and so on. Also care about the vitamin B and C content is at a very high level. Chlorophyll, a major component in plants, is used extensively to provide a green color in almost every which way. Carotenoids, the various colors in orange and red, are also utilized, as are certain anthocyanins (red, blue and purple). Different levels of pigmentation because varying levels of coloration between fruits and vegetables, allowing for certain cultivars of fruit and vegetables to look very different from others. Antioxidants act as carotenoids and anti-cyanins in the body. While chlorophyll does not appear helpful to human physiology, chlorophyll-rich foods are also high in 13 carotenes. I have 13-carotene yellow-orange blocks of green chlorophyll. Oe-carotene, 13-carotene, 13-cryptoxanthin, and lycopeneare the most widely researched carotenosides in human health.

Red may have an anthocyanin or a lycopene relationship, for example, and so on. Studies show that white foods, including garlic and onions, have allyl sulphides that impart that white garlic/onion flavor, but these sulfides are not found in other white vegetables, including apples, pears, deer, piglets or snowshoes. Chlorophyll-containing fruits and vegetables such as plums, blueberries, daikon radishes, lettuces, eggplants, arugula, spinach, and green tomatoes do not share the same chemical profiles as other food additives such as vitamin C, minerals and phenolic acids. Foods high in carotenoids include spinach, beets, radishes, kale, apricots, cauliflower, carrots, and chili peppers. They might also include natural spices derived from chili peppers such as capsaicin, and some of these do not contain glucosinolates (iceberg lettuce, green peas, and green beans). The peel of many fruits and vegetables varies in color which can vary from the bottom of the fabric. The outer layer of fruit or vegetables is generally a very small proportion of the total mass, and because of that, it is impossible to eat. The color of a half-peeled banana can often be inaccurately determined to indicate the consistency of the banana's internal food stuff. More specifically, looking from the outside in, examples include: summer squash yellow or green peel; cucumber green peel; aubergine purple peel; potato red peel; and apple red peel, green and yellow peel. Another difficulty in the analysis of fruit and vegetable pigment data is that there is a variety of cultivars, and their concentration of color and pigment can vary. As an example, several properties, such as color and splendor, can be observed in the cherry variety (Zhang, 2017).

There are various carotenes in varying quantities in sweet potatoes, including red, white, yellow, and orange varieties. The grading, grading, and preparation requirements of the classification of fruit and vegetables shall be according to the concentration of the food ingredients and by reference to the processing requirements of the operation of the classification of fruit and vegetables. Before the program, the section contained raw, whole, juiced, cooked, frozen and dried whole fruit and fresh, pureed, cooked and frozen whole vegetables. There have been many other ways to preserve foods and other pasturage beyond the use of ice and vegetable canning and freezing of foods. Conversely, suppliers and producers have added vitamins and folacin to these pasturages which have not increased the nutrition value of these foods but have dramatically increased the cost of these subsidized pasturages. Unlike the other products of the market, they are also different as either raw materials or cultivars used for freezing and canning. Just like the differences in the traditional tobacco leaf, there are chemical differences introduced during the manufacturing process that greatly affects the product contained within the nutshell. Many ways are used, such as drying fruit and vegetables. And the other thing that was done is that those products were reduced in weight by removing moisture, and if they did get water, it was removed.

- Other ingredients that can be used should be weighed on a weight basis and
- The pulp and fiber should be removed from fruit and vegetable juices.

The fruit and vegetable mix used to make juice may vary, as some companies purchase slightly different varieties from the supermarket than the ones sold on the market. Commercial oranges and juice are often fortified with calcium. This probably means that the more unfortified juice will have a marginally higher amount of unfortified juice. Fruit and vegetable juices and dried fruits contain differing amounts of nutrients and bioactive ingredients on a body weight basis relative to their fresh, frozen and fried counterparts. These fruits and vegetables may help protect against cancer. For example, the orange ingredients may consist of two to three oranges and the dried prunes may weigh at least less than one fresh orange. These results show that there can be a major difference in the consumption (weight) of dark green leafy vegetables between the raw and the cooked vegetables. An example of this is that when you cook raw spinach, the volume is reduced to half the raw volume. The processing and preparation of fruit and vegetables that have been peeled and other ingredients (such as fat, sugar and soup), used to prepare salads and made with soft scrambled eggs, cream or butter, mayonnaise, pickled vegetables and mashed potatoes, tomatoes and greens, additional sugar syrup, fruit juice or lemon juice, or made with finely chopped, soft fried or spun dough.

Processing

The classification of fruit and vegetables by processing and preparation methods can be of specific importance to populations using a limited variety of local plants and processing techniques to change their composition and minimize the intake of important food products. Because varieties of fruit-bearing and vegetable-bearing plants grow in different parts of the world without being restricted to certain processing methods, the classification procedures of these crops will likely not be useful on a global level. Concern about the growing trend of eating junk food has prompted scientists to develop dietary approaches for epidemiological studies whose aim is to obtain scientifically designed data on the content of different types of food. As you know, as part of that research, investigators develop a system that allows them to determine the average intake of fruit and vegetables in different countries. This paragraph describes dietary assessment methods, such as food frequency questionnaires, which relate the meaning and classification aspects of fruit, healthy and vegetables. The diets of people with disease say different things about fruit and veg, depending on whether a study is designed to measure dietary patterns or to find one particular disease-causing factors in fruit and veg.

Open-ended recordings of food records from the end to the beginning of the week/day/month, like a 24 hour recall, provide continuity in the identification and measurement of fruit and vegetable diets, like inspectors who organize the results as necessary after the detailed analysis has been completed. The questionnaire with questions on how to assess individual foods, how to group foods, and how to categorize foods, was carefully prepared, and the preferential choices and categories are very clear (FFQ). It is popular to differentiate between the consumption of fruit and vegetables with the consumption of sugar, fats, protein and alcohol. There's a difference in the number of fruits and vegetables on the list; what fruits and vegetables are supposed to be foods, and what plants are considered to be; grouping certain fruits and vegetables into other types of food, and there's also a difference in how they present and display their subject. For example, if people used oranges as the main energy source, the oranges would not be classified as fruits, that is, the main energy source. You may also see foods like peanuts, rice, white potatoes, and plenty of candy also served. Soybean products typically make up the main meat of the goods, but soya beans are commonly found there (tofu, miso, temph, analogues of soy-based moat, soymilk). Typically, Food Frequency questionnaires ask about the occurrence of mixed foods containing fruit or vegetables separately from fruit and vegetables (casseroles, stews, stir-fries; spaghetti, rice and pizza with tomato sauce; soups with vegetables; and pies containing fruit, pumpkin or sweet potato). Decisions on the place and description of fruit and vegetable ingredients in FFQs can be made on the basis of their common use in dietary habits. Learn about some global popular cuisine from around the world, including tomato ketchup, paste, purée, sauce and salsa; fried potatoes, tomatoes, peas or other vegetable soup, garlic and onion, coconut and sauerkraut, beef and olives. Usually, FFQs do not accept potato chips, sauces, jellies, jams and candied fruit as vegetables or berries. As meal guidelines and associated food suggestions for children and youth are provided in schools and the general population by public health and/or agro-medical organizations and health care professionals (dietarians, nurses, physicians), the food groups presented in the Dietary Guidance are likely to be recognized to many of the participants in the sample (Zhang, 2017).

2.5 ADVANTAGES OF FRUITS

The fruit is an excellent food supply with a high nutrient value for essential vitamins. Fruit, like flavonoids, can have a wide variety of well-being antioxidants.

Individuals with heart disease, cancer, inflammation and diabetes decrease the risk of eating a diet high in fruits and vegetables. Citrus fruit and berries can be highly powerful in the prevention of diseases.

As a powerhouse, fruit and vegetables, the 2014 study assessed both high nutrient density and low calories. Lemons came from the top of the map, followed by plum, peach, lime and grapefruit, pink or rose.

See the diet and various health benefits of these fruits and other fruits in the grocery store in this article.

Citrus,

Citrus fruit contains vitamin C and other useful wellness antioxidants.

Lemons are citrus fruits, which people often use in conventional medicines due to their therapeutic effects. They contain vitamin C and other antioxidants, comparable to other citrus fruits.

Human well-being requires antioxidants. These compounds produce free radicals in the body that can kill the cells of the body and lead to cancer diseases.

Researchers believe that flavonoids have antibacterial, anticancer and antidiabetic effects on lemon and other citrus fruits.

Citrus fruits, like lemons, are phytochemicals, which are medicinally important active ingredients. They shall include the following:

- Vitamin C.
- Folic acid.
- Pectic Acid.
- 1 48 g citrus juice contains grams (g) or milligrams (mg) of nutrients:
- Eleven calories.
- 3.31 grams of carbohydrate
- Dust at 49 mg.
- 18.6 mg vitamin C

- Calcium 3 mg
- 0.1 g of fiber

Lemons are also found in the following products: thiamine, riboflavin, niacin, vitamin B-6, folate and vitamin A.

How to have these orange berries

For a seasoning of drinking water, use lemon juices or splash over salads or seafood. To relieve the sore throat, consider applying a teaspoon of honey to the hot water. They also eat organic lemon rinds. Some people use rinds in recipes.

Strawberry

Strawberries are ripe red fruits with high water content. Seeds are full of nutritional fiber per meal. Strawberries contain a wide range of stable minerals and vitamins.

This includes anthocyanins of particular value that are flavonoids that can improve the defense of the heart. Fat and potassium in strawberries can also help a healthier heart.

In one study, the odds of heart disease for women who eat three or more strawberries and peaches a week are lower than those who have lesser intakes.

Strawberries and other live berries are also known as flavonoid quercetin. It is a natural anti-inflammation compound.

- Serve 3 large strawberries containing the following nutrients:
- Seventeen calories in size
- 4.15 g of carbon dioxide
- 2 g of fiber
- Calcium 9 mg
- 7 mg of magnesium
- 83 mg potassium
- 31.8 mg vitamin C

Strawberries also contain thiamines, riboflavin, niacins, folate and vitamins B-6, A and K.

Strawberries are multipurpose herbs. People can eat it raw or add it to the cereal or tea or yogurt, mix it in a smoothie or jelly.

Oranges

Oranges are sweet, sweet, mineral and vitamin fruits.

Oranges are one of the richest sources of vitamin C, with medium-sized fruit supplying 117 percent of human vitamin C daily.

- 141 g of orange also contains the following nutrients:
- 25 Calories
- 16.27 grams of hydrocarbons
- Fiber 3.4 g
- 61 mg of calcium
- 14 mg of magnesium
- 238 mg of potassium
- 63.5 mg vitamin C

Vitamin C is the body's effective antioxidant. This vitamin is also essential to the activity of the immune system. This improves the immune system by encouraging the body to absorb iron from herbal food.

The human body cannot produce vitamin C on its own, so people have to get this vitamin out of their diet. In addition, oranges contain high amounts of pectin, a fiber that helps to keep the colon healthy by binding and removing toxins that cause cancer in the colon.

In oranges, the following healthy vitamins are also present:

Vitamin A, essential for healthy eyes and skin,

B-vitamins such as thiamine and folate help facilitate the healthy development of the nervous, reproductive and red blood cells.

People should eat their own orange as a refreshing snack or enjoy a bottle of pure orange juice. Take orange juices at home, or choose a brand of fresh juice, the name of which does not stand out.

In salads, yogurt or as a cereal roof, visitors can also add flavor to orange peel.

Limes

Lime is a sweet citrus fruit with a range of well-being advantages.

Limes, like other citrus fruits, provide a healthy dose of vitamin C. They have the same health benefits, antibacterial and antioxidant impact.

One lime juice is needed for the following nutrients:

- Eleven calories.
- 3.7 grams of starch
- 6 grams of calcium consumption
- 4 mg of magnesium
- 51 mg potassium
- 13.2 mg vitamin C

Lime is doing well in the savory berries. Treat your spicy salad dressings or dishes with lime juice or rice. If you do not, use hot or cold water and lime juice to make a soft drink.

Grape fruit

Grapefruit contains flavonoids that can help protect against some infections, asthma and obesity.

Grapefruits are bitter fruits with the well-being of minerals and vitamins. Pink, red or white grapefruits can be available.

Half of the grapefruit contains the following nutrients:

- 52 calories per dish
- 13.11 g Carbohydrate
- 2.0 g of fiber
- 27 grams of calcium intake
- 11 grams of magnesium
- 166 g potassium
- 38.4 grams of vitamin C

Flavonoids can be used in grapefruit to guard against some cancers, inflammation and obesity.

Analysis has shown that grapefruit compounds called furanocoumarins may help defend them from oxidative stress and tumors and promote healthy bones.

Data from this study suggest that the effects of pomegranate furanocoumarins on breast, leukemia and skin cancer can be particularly beneficial. Further tests on animals and humans are required to verify the properties. People will have to see a doctor before adding grapefruit to their diet, since they may clash with other medications.

Try adding slices of grapefruit to a fruit salad or squeezing the juice into the water to make a drink. Otherwise, the individual can buy pure grapefruit juice from a supermarket.

Chapter 3

Importance of Fruit Production

CONTENTS

3.1 INTRODUCTION

The intensive growing of perennial plants, the fruits of which have economic significance, is the subject of fruit and nut production (a nut is a fruit, botanically). It is one part of the broad horticultural theme, which also includes the cultivation of vegetables and the development of ornamentals and flowers. In general terms, botanists describe a fruit as the fleshy or dry, ripened ovary covering a plant's seed. A pomologist, or expert in the practice and research of fruit growing, describes it much more broadly as the fleshy edible portion of a perennial plant related to the growth of the flower. A nut is any seed or fruit which consists of a kernel surrounded by a hard or brittle shell, typically oily. Most edible nuts are well known as dessert nuts, e.g., almond, walnut, cashew, pecan, pistachio, etc. The nuts are not always nutritious. Others may be known as oil seeds, used as sources of oil or fat; others are used for ornamentation. Based on characteristics of shape and form (morphology), the botanical concept of a nut is more restrictive: a rough, dry, single-celled, single-seeded fruit that does not break open at maturity (Valpuesta, 2002).

Figure 3.1: Fruit is a food commonly consumed that is deemed nutritious and essential for a balanced diet. The most grown varieties of fruit worldwide are watermelons and bananas.

Source: https://www.statisticbrain.com/fruit-production-and-consumption-statistics/

One of the significant and age-old activities practiced in India since ancient times is fruit growing. The cultivation of fruit crops plays a key role in humanity and the nation's overall status. The quality of life of the citizens of a nation depends on fruit production and consumption per capita. There are economic and nutritional benefits to fruit growing. The production of fruit crops plays a significant role in any nation's prosperity. It is widely reported that people's quality of living can be determined by the consumption and production of fruits per capita. Fruit crops can yield a higher tonnage per unit area than other crops in the field. For example, for two crops per year, a wheat crop produces an average of 12-15 tonnes from an area of one hectare. Even a hybrid rice variety can only yield a total of 24 tons per hectare of land for three crops per year, while a banana crop can yield 35-40 tons per hectare. In 2 1/2 years, papaya yields 100-150 tons per hectare, which is 40-60 tons per year, and mango yields 25 tons per hectare. An amount of 60-80 tons per hectare in two harvests per year can be produced from grapevine in a tropical climate.

It is found that fruits are a rich source of vitamins and minerals. The essential component of beta carotene, which is basically the precursor of vitamin A, is mango, papaya, and jack, for instance. Quite strong sources of readily available beta-carotene, 1990 ug per 100 g in mango and 880 ug per 100 g in papaya have been estimated to be mango and papaya fruits. Though bajra only supplies 132 ug of beta-carotene per 100 g, wheat hardly supplies 64 ug per 100 g of beta-carotene. It is heartbreaking to think that raw rice that is the main energy source in the South Indian diet does not possess any carotene at all.

Many phytochemicals present in fruits function as strong antioxidants that protect muscles and tissues from damage caused by free radicals, mitigating their harmful effects, according to recent research findings. They are the bioactive plant substances that provide them with color, taste, odor, and defense against not only plant-related diseases, but also human diseases. As a result, hundreds of such plant compounds are currently being studied for their function in cancer and other degenerative disease prevention. Bioflavonoids (Vitamin P), glucosinolate, antioxidant vitamins (C and E), carotenoids, lycopene, and phenolics, are some of the promising phytochemicals acting as antioxidants. Besides being the main sources of vitamin C and folate, limes, lemons, oranges, and grapefruits are abundant in a class of phytochemicals called limonoids. It has been discovered that this antioxidant is amazingly effective against cancer (Valpuesta, 2002).

The most popular food preferred for a patient suffering from extremely high fever is sweet orange. As well as being readily assimilated, it has a cooling effect. A range of banana fruit, Peyan, is offered to patients with chickenpox as it lowers the body's elevated temperature. It is worth noting the potential of fruit crops for the economic growth of the country. The country's stability lies in building up its reserves for foreign exchange. As a country with complex climatic conditions varying from subtropical to tropical to temperate, India has enormous potential for the development and export of various fruits. During 1998-99, 1.18 lakh tonnes of fruits and nuts priced at Rs.24,714 lakhs were shipped from India. With 48.57 million tonnes, India ranked second in annual fruit output during 2002, accounting for around 10.3 percent of total world fruit production. Recent policies by the Government of India to facilitate the export of fruits and their goods by announcing concessions to the fruit industry, such as decreased air freight charges and exclusion from storage charges for frozen air cargo at international airports, have prompted several private entrepreneurs/ corporate organizations and NRIs to enter a larger area of fruit crops with a wider area of planting.

With a view to emphasizing the growing importance of the fruit industry, the above account is given. As can be seen from the information provided in it, the fruit culture, from the viewpoint of increased agricultural productivity, nutrition, trade, and fruit-based industries, is vital to the nation's health and economy. This is focused on India; however, the concept is broadly seen in many other countries all over the world (FAO, 2020).

3.2 NUTRITIONAL VALUE

Eating fruit has health benefits, and individuals who consume more fruits and vegetables are likely to have a decreased risk of certain chronic diseases as part of an overall balanced diet. Fruits contain nutrients essential to your body's health and maintenance. They are put on the crest of our edibles by the nutrient value of fruits. In large amounts, fruits contain minerals and vitamins. Fruits are humanity's oldest meal. Everyday fruit-taking enhances our vitality. Nutrition experts recommend for us to eat at least 115 grams of fruit daily for a healthy diet.

Vitamins A, B and C are abundant in fruits such as papaya, mango, guava, jackfruit, pineapple, lemon, etc. They have potassium, iron, magnesium, and calcium, in addition. Adequate quantities of phosphorus and amino acid are found in wood apples, litchi, almonds, karamcha, etc. In India, rice and wheat are food staples. People consume fruit as their staple food in many countries. For example: People in the countries of South America eat banana as their main meal course. Our food shortage can be mostly minimized by banana, jackfruit, guava, pineapple, etc. As food, fruits are important because they have adequate quantities of vitamins and minerals. We need to consume 115 grams of fruit per day. Fruits increase our capacity of digestion. A daily consumption of fruit makes us breathable and hearty (T. Radha & Lila Mathew, 2007).

Citrus fruits are high in vitamin C, with Barbados cherry being the richest source of vitamin C. The richest source of fat (64.5 percent) is walnut, which contains the highest calories per edible portion unit. The richest source of 21.2 percent protein and 630mg/100g of thiamine is cashew nut. The best source of niacin, phosphorus, and calcium is Dry Apricot. The richest source of iron (39.1%) is dry karonda, followed by date palm (10.6 percent). Litchi has been found to be high in calcium. The richest source of riboflavin is Bael.

Figure 3.2: A fabulous food is a fruit. It is rich in nutrients; it is easy, and it can be inexpensive. Fresh fruit is compact, and it normally needs no refrigeration if it is whole. Normally, most fruits are low in fat, sodium, and calories and are a source of many essential nutrients.

Source: https://www.medindia.net/amp/patients/lifestyleandwellness/declining-nutritional-values-of-fruits-and-vegetables.htm

Proper nutrition encourages children's optimum development and growth. Healthy eating helps to avoid elevated cholesterol and high blood pressure and helps to minimize the risk of chronic conditions such as heart disease, cancer, and diabetes. Healthy eating helps in the reduction of obesity, osteoporosis, dental caries (cavities) and iron deficiency. A poor diet can lead to an excess of energy (e.g., consuming more calories than one spends through physical activity) and can raise one's risk of being overweight and obese. The risk of cancer of the lung, esophagus, prostate, colorectal, and stomach, may be increased by a poor diet (T. Radha & Lila Mathew, 2007).

A large range of nutritious food is available in the USA. Many Americans, however, do not eat the variety of foods that include all the nutrients needed while remaining within the calorie requirements. The consumption of milk, whole grains, fruits, vegetables and milk products, and oils in the United States is lower than recommended. Therefore, dietary levels of certain nutrients—vitamin D, calcium, dietary fiber, and potassium, dietary fiber—are poor enough for both adults and children to be of public health concern. There are high and rich vitamin and mineral contents in tropical and sub-tropical fruits that can form an integral part of a nutritionally healthy diet. Since fruits appear to contain significant amounts of potassium, phosphorus, calcium, and sometimes iron and magnesium, they are especially essential for providing children with the building blocks of healthy muscles, bones, teeth, and brain, as well as for promoting a fully functional nervous system, cellular metabolism, and protein digestion.

Some fruits are exceptionally rich in dietary fiber, such as acai fruit, which helps digestion and lowers blood cholesterol. There are high calorific levels for some fruits. For example, peach palm pulp is not overly sweet and is comparable to, or exceeds, certain carbohydrates, fat, and energy cereals. The dried pulp of açai berry and safou, likewise outperform cereal crops when fat and energy contents are compared. Some fruits are also a good source of protein, with both important and non-essential amino acids found in the peach palm fruit. In several different ways, fruit can be integrated into diets. On their own or in desserts, ripe fruits are always sweet and eaten fresh. By chilling, the flavor of certain fruits such as mangosteen and custard apple can be enhanced. It is also possible to store the fruit in jams and pickles, or to dry and powder it for use in drinks or sweets.

Immature green fruit can be cooked and used like vegetables in many species, such as jackfruit. It is possible to fry and eat seeds of certain species as snacks and can be a good source of oil. In addition to giving nutritious side dishes and desserts for the entire family, by offering readily available snacks, fruit trees planted near the house can be especially valuable for young kids who have particular nutritional and high energy requirements (T. Radha & Lila Mathew, 2007).

Regarding the nutrients:

- Most fruits are low in calories, sodium, and fat, naturally. None of them got cholesterol.
- Fruits are the source of many under-consumed important vitamins, including folate (folic acid), vitamin C, dietary fiber, and potassium.
- Potassium-rich diets can help keep blood pressure stable. Bananas, prune juice, and plums, apricots and dried peaches, orange juice, honeydew melon, and cantaloupe are fruit sources of potassium.
- As part of an overall balanced diet, dietary fiber from fruits helps reduce blood cholesterol levels and can reduce the risk of heart disease. For proper bowel function, fiber is essential. This tends to decrease diverticulosis and constipation. Fiber-containing foods such as fruits help with less calories and provide a sense of fullness. Cut-up or whole fruits are dietary fiber sources; there is no or little fiber in fruit juices.
- Vitamin C is necessary for the growth and repair of all body problems, helps to heal cuts and wounds, and maintains healthy teeth and gums.
- Folic acid (folate) allows red blood cells to develop in the body. Adequate folate from foods should be eaten by women of childbearing age who can become pregnant, plus 400 mcg of synthetic folic acid from fortified food and supplements. This decreases the risk throughout fetal development of neural tube defects, spina bifida, and anencephaly.

Some of the health benefits include:

- As a bulk laxative, dietary fiber is known to work. It prevents the absorption of LDL cholesterol in the intestine, helps control the motions of the bowl and helps protect the mucous membrane of the colon from chemicals causing cancer. The fiber helps to improve digestive health and reduce issues with constipation. The soluble fiber will help maintain the levels of blood sugar stable. For a longer period, fiber may keep the stomach full and decrease the need or urge to consume more food, aiding in weight loss. It helps in the reduction of diabetes, obesity, and blood cholesterol.
- Vitamin-C is a natural antioxidant that is strong. It protects against oxidative stress and improves the role of the immune system. For

all body tissues, it is essential for growth and repair. It is needed within the body for collagen synthesis. In the human body, collagen is the primary structural protein necessary to preserve the integrity of blood vessels, skin, organs, and bones. It improves immunity to defend against common diseases such as flu, cold, cough, and to prevent and treat anemia effectively. It helps in healing wounds and cuts, keeping gums and teeth healthy.

- For normal eye vision, proper functioning of the immune system, reproduction, and maintenance of healthy soft and skeletal tissue, teeth, skin, and mucus membranes, vitamin-A is essential. It helps to prevent cancers of the oral cavities, liver, and lungs.
- Vitamin-E prevents free radicals from destroying fatty acids and cell membranes, allowing the body to combat symptoms of skin aging, such as loss of elasticity and wrinkles. It is famous for sex drive ties.
- For several coagulant factors in the blood and in bone metabolism, vitamin-K is crucial.
- B-complex vitamin groups such as pyridoxine (vitamin B6), folate, pantothenic acid, niacin, thiamin, and riboflavin serve as cofactors that enable the body to metabolize fats, proteins, and carbohydrates. Such vitamins can help to control the risk of heart disease. Folate (folic acid) allows red blood cells to develop in the body.
- Nerve cells are covered by antioxidants due to oxidative stress, which eventually reduces the risk of neurodegenerative diseases such as Alzheimer's and Parkinsons. Reductions in heart disease have been associated with polyphenolic antioxidants. There are anti-infective, anti-hemorrhagic and anti-inflammatory effects of an antioxidant known as tannin.
- Antioxidant flavonoids such as β-carotene, crypto-xanthin, and lutein-zeaxanthin can protect cells and protecting them from pancreatic, liver, lung, endometrial, breast, prostate and colon cancers. The risk of cataracts and macular degeneration is decreased by Lutein. The healing mechanism of oral ulcers and peptic is accelerated by the detoxifying effects of flavonoids.
- Carotene helps guard against cancers of the lung and oral cavity. Carotenoids and xanthophylls aid eyesight against damage associated with aging.

- Lycopene prevents UV rays from destroying the skin and provides protection against prostate cancer.
- The oxygen-carrying potential of the blood is determined by iron being a part of hemoglobin within the red blood cells. Iron is required both for the formation of red blood cells and for cellular oxidation.
- Potassium helps control heart rate and blood pressure as an essential part of cells and body fluids. This also decreases the risk of developing kidney stones and significantly reduces the loss of bone.
- Calcium is a crucial part of the conduction of bones, teeth, muscle contraction, blood clotting, and nerve impulses.
- As a co-factor for antioxidant enzymes, the body uses manganese.
- For bone metabolism and for the development of white and red blood cells, copper is necessary.
- For bone growth and its reinforcement, magnesium is important. It also has a cardiac-protective function.
- Omega-3 and 6 fatty acids can help lower cholesterol levels in the blood, control blood pressure, and reduce the risk of heart attacks and disease.

Figure 3.3: Above are some of the main benefits of fruit including lemons, watermelon, oranges, grapes, apples, bananas, and pineapples.

Source: https://www.pinterest.com/pin/118501033920221475/

3.3 MEDICINAL VALUE

Eating a fruit-rich diet as part of an overall healthy diet can decrease the risk of many types of illness. Here are the advantages of various kinds of fruit for human health:

a. *Apple:* The fruit helps reduce levels of cholesterol in the body. The risk of skin disorders, improving digestion and detoxifying the liver is decreased by an apple a day. Kidney stone formation is prevented by Apple cider vinegar used as a beverage. Halitosis, neuritis, insomnia, anemia, tuberculosis, gonorrhea, bronchial asthma, gall bladder stones, obesity, and arthritis have been recommended for apples. It helps to prevent diarrhea, constipation and neutralize irritable bowel syndromes. It helps strengthen the immune system, regulate weight, maintain a healthy heart, prevent cataracts, and prevent cancers of all kinds.
b. *Apricot:* Apricot helps to enhance normal vision, proper functioning of the immune system, reproduction, and soft and skeletal tissue, teeth, and healthy skin maintenance. Apricots are a major source of potassium that helps keep appropriate fluid balance, maintains muscle function, controls heartbeat, promotes safe digestion and strong bones. Apricots contain large amounts of fiber that is both soluble and insoluble, supporting safe levels of blood glucose and cholesterol. Antioxidants in the apricot serve as defensive scavengers against free radicals produced from oxygen and reactive oxygen species that play a role in ageing, cancers, and different processes of disease. It helps shield the individual from cancers of the lung and oral cavity.
c. *Avocado:* Avocado can assist in controlling the risk of heart disease. In avocados, soluble fiber can help preserve stable blood sugar levels, assist digestion, and help maintain regularity. The high level of potassium in the fruit will help regulate blood pressure. Avocados are quite high in mono- and polyunsaturated omega-3 fatty acids, which can help lower blood cholesterol levels and lower the risk of heart disease. Avocado may decrease the risk of macular and cataracts degeneration. It has strong antioxidants related to the immune health. It has been shown that avocados minimize the risk of some cancers, including prostate, skin, and mouth cancers.
d. *Bael fruits:* A species native to India is Bael fruit, also known as the "Wood Apple". Bael fruit is used to make the popular drink known as sherbet. Since 2000 BC, it has been renowned for its medicinal values. It offers relief from dysentery, diarrhea, cholera, hemor-

rhoids, breathing problems, scurvy, piles, peptic ulcer, indigestion, and constipation. Bael fruit helps. It also strengthens the immune system, battles against viral and bacterial infections, decreases inflammation, prevents cancer, and other inflammatory conditions. It stimulates the production of milk for nursing mothers, treats diabetes, and improves eye health. It helps to prevent, treat heart disorders, and regulate cholesterol from various sexual dysfunctions.

e. *Dates:* Fresh dates consist of basic sugars like fructose and dextrose that are readily digestible. The fruit is rich in dietary fiber that inhibits the absorption of LDL cholesterol in the gut. The fruit helps to prevent the mucous membrane of the colon from toxins that cause cancer. The fruit has anti-infective, anti-hemorrhagic and anti-inflammatory properties. It assists with vision and helps to protect against cancers of the oral and lung cavity. The fruit's antioxidants have the potential to protect cells. It allows the body to metabolize the digestion of carbohydrates, proteins, fats, and bones. Dates are an excellent source of iron, which defines the oxygen-carrying ability of the blood as a portion of hemoglobin within red blood cells. They are excellent sources of potassium, which helps to control blood pressure and heart rate. Calcium is an integral component of the conduction of nerve impulses, blood clotting, muscle contraction, teeth, and bone.

f. *Figs:* An important source of antioxidants, vitamins, minerals, and dietary fiber are both fresh and dried fig fruits. Fig fruit is high in dietary fiber, which is amazingly effective for obesity and helps minimize cholesterol in the blood. Fig fruit helps scavenge free radicals extracted from the body from toxic oxygen and thereby protects us from tumors, diabetes, degenerative diseases, and infections. It helps to regulate blood-glucose levels and lower blood sugar levels in type II diabetes mellitus. The B- complex group vitamins operated as co-factors for fat, protein, and carbohydrate metabolism. This increases the formation of red blood cells for cellular oxidation as well. In figs, tryptophan causes healthy sleep and helps to get rid of sleep problems like insomnia. To treat sexual weakness, figs have also been used. The fruit is especially useful for healing of hemorrhoids and treating chronic constipation. It helps to increase the density of bones and prevent the thinning of bones (osteoporosis). This significantly reduces the risk of prostate cancer, cancer of the breast and cancer of the colon. Fatty acids such as potassium, omega 3 and 6 in fig fruit help maintain blood pressure and prevent coronary heart attacks. Figs can decrease fatigue, boost memory in the brain and avoid anemia. Figs are abundant in natural benzaldehyde and anti-tumor agents, which ensure that the human body can destroy viruses, fungi, and bacteria.

g. *Grapes:* In many cultures, grapes have been known as "the queen of fruits" for centuries. Phyto-chemical and antioxidants in grapes with anti-cancer, anti-microbial, anti-inflammatory, and anti-allergic activity have been reported. Grapes play a defensive function against cancers of the prostate and colon, degenerative heart disease, and coronary heart disease. Grapes help to prevent stomach irritation, to cure indigestion, and avoid dyspepsia. The latest research says that Concord grape juice, purple in color, helps to avoid breast cancer. Grapes help reduce cholesterol levels and are extremely healthy for eye vision. Grapes help to minimize the acidity of uric acid and attempt to expel acid from the environment, resulting in a decrease in kidney pressure. Grapes tend to increase the level of humidity in the lungs and are also known to be of high medicinal benefit for asthma. In overcoming constipation, grapes are very effective. It lightens scars, cures uneven skin tone, rejuvenates the skin, and offers protection against sunburns.

h. *Guava:* Guava is a tropical fruit rich in nutrients with a high profile. The fruit easily falls into the modern functional food category, also referred to as a 'super-fruit' with its unusual flavor, taste, and health-promoting qualities. Guava has a low fat and calorie content, but it contains essential flavonoids, poly-phenolic, minerals, and vitamins, compounds that play a key role in anti-aging and cancer prevention. The fruit has the largest source of Vitamin-C that defends oxidative stress and improves immune system function. For the synthesis of collagen, vitamin C was required to preserve the integrity of blood vessels, skin, organs, and bones. It contributes to the preservation and defense of healthy mucus membranes, skin, and cancers of the oral and lung cavity. In pink guavas, lycopene inhibits skin damage from UV rays and provides prostate cancer protection. Soluble dietary fiber in guava helps in protecting the mucous membrane of the colon through the reduction of toxin exposure time as well as binding to colon chemicals that cause cancer. Potassium in guava is an essential component of cell and body fluid that is needed for the development of red blood cells to help regulate copper, blood pressure, and heart rate.

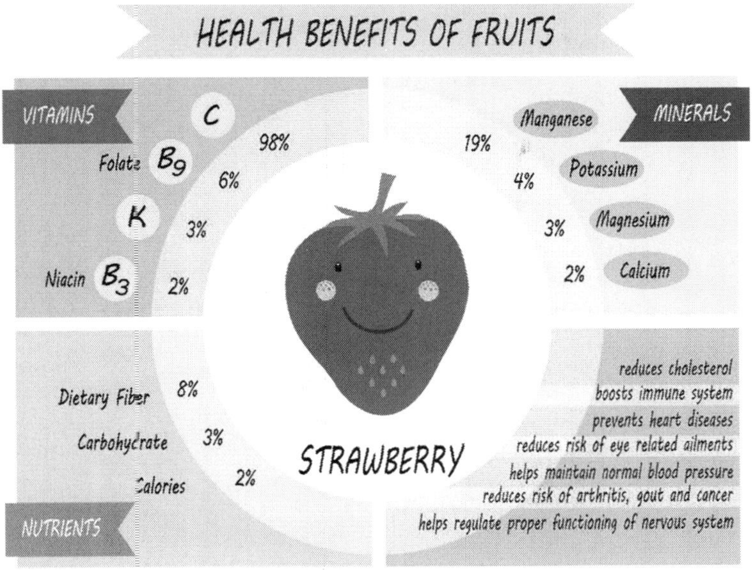

Figure 3.4: Above are some of the benefits of eating strawberries. Regarding vitamins, strawberries have a high percentage of vitamin C. Other vitamins include vitamin B9, K, and B3.

Source: https://www.dreamstime.com/stock-illustration-cute-infographic-page-health-benefits-fruits-like-vitamins-minerals-nutrients-image94319030

i. *Kiwi fruit:* There are many health-promoting dietary fibers, vitamins, minerals, and antioxidants in kiwi fruits. The fiber content helps to protect the mucous membrane of the colon by reducing the duration of exposure to toxins and binding to chemicals in the colon that cause cancer. Constipation and other digestive disorders are avoided. The fruit is an outstanding vitamin C source that allows the body to establish a resistance to infectious agents and scavenge free radicals that are harmful. It has also identified a role in reducing neuronal damage to the brain in Alzheimer's disease patients. The fruit acts as an aspirin-like blood thinner, so it helps avoid clot formation within the blood vessels and protects against stroke and the risk of a heart attack. Kiwi-fruit seeds are an excellent source of omega-3 fatty acids, decreasing the risk of stroke and coronary artery disease. The fruit helps control blood pressure and heart rate. For bone-reinforcement, magnesium in the fruit is essential.

j. *Lemon:* The lemon's citric acid is a natural preservative that helps to digest smoothly and helps remove kidney stones. Lemon ascorbic acid is effective in preventing scurvy. It allows the body to safeguard against cancers of the lung and oral cavity. It helps to regulate the heart rate, blood pressure, nausea, and dizziness. Lemon has coagulant and antiseptic properties, so internal bleeding can be prevented. Lemon juice can help reduce cholesterol levels when used with honey and lukewarm water. It helps to ease respiratory and breathing issues and to relieve asthma attacks. It is possible to cure diseases such as cholera and malaria and to serve as a blood purifier. Lemon is an antiseptic and aromatic agent and is helpful for foot relief. Lemon can treat arthritis and rheumatism and is a diuretic. Lemon is an outstanding fruit that battles against issues linked to infections of the throat. To relieve mental stress and depression, it is widely used.

k. *Litchi:* Litchi fruit helps to increase blood flow in tissues, minimize weight and shield the skin from harmful UV rays. It seeks to enhance the body 's immune system and to scavenge harmful free radicals that are pro-inflammatory. It helps in metabolize fats, protein, and carbohydrates, regulating blood pressure and heart rate while it provides protection against stroke and coronary heart diseases. Litchi has anti-cancer properties and helps to combat diseases that are lethal and fatal. Litchi maintains strong digestion, maintains a healthy stomach, enhances appetite, and cures the stomach's heartburn and burning feeling. This preserves healthy bones.

l. *Mandarin:* Mandarin can reduce the risk of cancer of the liver and help prevent breast cancer. The antioxidants contained in mandarins disarm free radicals and prevent cellular harm. It reduces the production in the body of bad cholesterol and promotes healthy cholesterol. It keeps the flow of blood steadily into the arteries, preserving regular blood pressure. It helps to control the immune system and body weight. It strengthens the healthy immune system and protects the skin against harsh UV rays, allowing the skin to withstand sun damage and free radicals. In the digestive and nervous system, mandarins avoid spasms. It is a natural purifier of blood that allows the body to flush out contaminants and unwanted substances. Daily mandarin consumption significantly enhances the complexion and provides skin that is radiant and blemish-free. Mandarin oil is excellent at developing new cells and tissues, so it helps to heal wounds more easily.

m. *Mango:* Mango is the nutritionally rich "king of the fruits" with unique qualities that promote color, aroma, taste, and health. Mango fruit protects against cancers of the colon, breast, leukemia, and prostate. High levels of fiber, pectin, and vitamin C help to minimize serum levels of cholesterol. Mangoes contribute to the regulation of heart rate and blood pressure. They clear blocked up pores and pimples while the extract from mangos enhances skin shine, encourage good eyesight, avoid dry eyes and night blindness. The fruit's tartaric, malic, and citric acids help to preserve the body's alkali reserve. Mango fruit can regulate the level of blood sugar and mango leaves help to normalize the level of insulin in the blood. Mangos are a great source of vitamin E, the common sex drive link. Digestion and removal are enhanced by the fiber in mangos. Mango juice could be used for heatstroke treatments. It stimulates the immune system and is strong and healthy.

n. *Orange:* Oranges are abundant in limonoids from citrus, which have been shown to help battle a variety of cancers, including skin, lung, liver, breast, stomach, and colon. Drinking daily orange juice reduces kidney failure and decreases the risk of stones in the kidneys. They are effective in preventing cholesterol build-up because oranges are full of soluble fiber. Oranges are full of potassium, which is the responsibility of an electrolyte mineral to help the heart function well. Oranges are full of vitamin C that, by neutralizing free radicals, protects cells. Chronic illnesses, including cancer and heart disease, are caused by free radicals. Oranges are full of dietary fiber that induces and relieves constipation by stimulating digestive juices. Oranges are high in vitamin A, which helps avoid and improve eye vision by avoiding macular degeneration. In oranges, magnesium helps to regulate blood pressure and to control high blood pressure. Oranges are full of β-carotene, a strong antioxidant that helps cells from free radicals and protects the skin.

o. *Pineapple.* Pineapples are a rich source of antioxidants that combat and prevent cell damage from free radicals. It helps protect against various cancers, arthritis, diseases of the heart, atherosclerosis, etc. Pineapples are abundant in vitamin-C that helps to effectively treat cough and cold. In pineapples, manganese helps to reinforce the connective tissue and bones. It helps to improve gums, to maintain healthy and strong teeth. Beta-carotene found in pineapples enhances vision. Pineapple is well-known for its anti-inflammatory properties, which in arthritis patients helps in preventing inflammation and giving the bones strength. To treat swelling, gout, sore throat and sinusitis, pineapples are useful. Pineapple antioxidants help to kill free radicals, prevent cancer, heart disease and lower cholesterol levels. Due to its powerful antioxidant influence, Vitamin-C in pineapple strengthens your immune system. Dietary fiber is found abundantly in digestion-promoting pineapples. In pineapples, higher concentrations of potassium and lower amounts of sodium help restore optimum levels of blood pressure. Intestinal worms are treated, and it relieves nausea. Eating pineapple is an excellent acne solution that causes the skin to appear younger and is a natural treatment for chapped lips, addresses hair loss, ideal for scalp disorder that is inflammatory.

3.4 IMPORTANCE OF FRUIT PRODUCTION IN SUSTAINABLE LIVELIHOODS

a. *Fruit and food security*: Food security ensures that specific households, from their own farms, can fulfil their daily food needs or have the means to purchase food from nearby off-farm sources. Fruit crops are an ideal addition to any small-scale farm and home and market garden since they need little cash and labor input once they are grown and do not need to interfere with staple crops if planted in suitable locations. Planting a variety of different species of fruit trees will provide a year-round supply of fruit for households to supplement other crops grown on the farm. Unlike other crops that may fail completely if subjected to climate stress, fruit trees grown that are admirably adapted to a region may have decreased yields but will typically survive one or two challenging seasons. Integrating fruit trees into farming systems can help provide the building blocks for the creation of sustainable agro-ecosystems by promoting biodiversity, improving soil fertility, and preventing soil erosion (Shri Mohan Jain & P.M. Priyadarshan, 2008).

Figure 3.5: Companies offer creative solutions every day to address the world's biggest problems, including environmental wellbeing, wellness, and equal and sustainable food systems.

Source: https://ccafs.cgiar.org/news/how-local-fruit-trees-contribute-food-security-and-resilience-climate-smart-village-daga-birame

b. *Fruits as a source of income*: A benefit of fruit is that, based on the household, they may be eaten or sold. Species such as custard apples are not perceived to be extremely commercial species and are often grown on a casual basis by smallholder farmers, integrating other farming activities seasonally. Their product is sold or eaten at home in local markets, adding variety to the local food basket and the income of the farmer. Increasingly, other plants are cultivated mainly for export. Thus, for example, the durability and two-week shelf-life of mangosteen fruit, coupled with its delicate taste and friendly appearance, contribute to growing exports to the People's Republic of North America, Europe, Taiwan, and China, from Southeast Asia. When trading fresh fruit, when costs are high, it is beneficial to grow varieties that fruit early or late in the season. The glut of abundance can cause prices to drop dramatically during the main fruiting season. Proper processing or storage will, in this case, significantly increase the income of a family. For instance, the price of jackfruit doubles or triples just a few months after harvest. For instance, the Philippines has seen a 100-fold increase in demand for jackfruit because of advancements in technologies that have enabled out-of-season sales of the fruit (fresh and processed). Some varieties, such as tree tomatoes, which are extensively grown in home gardens in South America, continuously fruit in subtropical areas offering a year-round source of income. Others, such as safou, play a vital seasonal role in family finances in Central and West Africa, by paying compensation at a time when farmers have few other sources of income but still need to pay for books, uniforms, and school fees. Small-scale processing will guarantee the availability of fruit tree production throughout the year, thereby improving food security and generating income for poor farmers (K. K. Jindal & R. C. Sharma (prof.), 2004).

c. *Sustainability for vulnerable groups and women*: It is considered that fruit tree crops are less labor-intensive than other crops. They need little weeding or tending once established and, if necessary, labor can be hired for a few days to assist with the harvest. Therefore, fruit trees may be especially suitable if labor is a limiting factor in a family, as might be the case when incredibly old people take care of incredibly young people or in households headed by women. In communities with a high prevalence of HIV/AIDS or when young men move to cities in search of jobs, these circumstances may arise. Fruit tree cultivation offers an opportunity to grow a regular food and especially income stream from the perspective of a poor, elderly or female farmer that can easily fit around other everyday household and farm activities, while offering reasonably low-risk returns. For individuals or households with restricted access to land, fruit trees are also ideal, as one or two trees can comfortably be grown around the farmhouse or even in an urban backyard. For women in societies where it is challenging for them to work outside the home, the place near the home may be important. It also ensures that the trees can be controlled without interfering with the domestic obligations of the woman and the fruit is readily available for food preparation.

The fact that food is regarded as the domain of women also makes fruit tree companies an appropriate activity for women to generate income. Small-scale manufacturing, such as solar drying, requires little investment in resources or technological expertise and can be an acceptable practice for individuals with little to no capital assets to incorporate into their daily routines. Community-based manufacturing businesses, such as washing or chopping fruit and filling containers, may provide jobs for disabled or elderly people at work.

3.5 ENVIRONMENTAL IMPORTANCE OF FRUIT PRODUCTION

The many advantages of growing fruit trees include their new, locally grown food yield. As another benefit, in urban and suburban environments, fruit trees grow well. Fruit trees help folks become linked to the growing process from a social point of view, while also providing a healthy source of food and food security. There are also several beneficial environmental advantages of planting fruit trees, from cleaner air to lowered energy costs and green jobs.

a. *Reduced CO_2 Emissions*: It is widely accepted that the burning of fossil fuels is the source of global warming. One of the biggest offending fossil fuels is carbon dioxide. Luckily, trees help compensate for the effects of CO_2 pollution. Trees need CO_2 to survive, including fruit trees. Trees act as air cleaners, or filters, removing CO_2 and ejecting into the atmosphere fresh oxygen. Planting a fruit tree or two in the garden landscape will do its part for the climate. One would also enjoy the advantages of fresh fruit, as well as reducing greenhouse gases. One acre of mature fruit trees will consume as much CO_2 as will be generated by driving 26,000 miles, according to TreePeople.org.

Figure 3.6: There are also several beneficial environmental advantages of planting fruit trees, from cleaner air to lowered energy costs and green jobs.

Source: https://www.phillyvoice.com/healthier-diets-better-environmental-impact-study-shows-food-groups-research/

b. *Reduction of energy costs*: In the home garden, fruit trees help save energy and reduce the costs of services, such as electricity and water. The need for cooling on the occasional hot day can be minimized by a fruit tree providing shade for your house. They lose less moisture by evaporation on sunny days when partial shade-loving plants are integrated near your fruit tree, eliminating the need for additional watering. Lawns shaded by fruit trees, since they do not lose moisture or get scorched by the light, are often healthier and colorful.
c. *Management of Storm Water*: Storm water drainage is also a concern in urban environments with few trees. While storm water appears harmless, runoff water absorbs and carries contaminants that travel to streams, rivers and, eventually, the ocean without proper absorption. On hillsides and cliffs, this runoff also contributes to erosion. By collecting some of the runoff and using it for hydration after a downpour, fruit trees help to mitigate some storm water management issues and erosion. If a yard is especially soggy, adding a tree is helpful in reducing problems in drainage.
d. *Green jobs*: A garden or orchard with fruit trees gives residents opportunities to learn about sustainable growth and grow their own food in a community environment. The orchards open the door for green employment and other small business prospects in some communities. In an economic sense, this not only benefits the economy but also helps the world by encouraging healthy living. The growing of fruits allows individuals to have a stronger connection to the growing process, as well as the knowledge of where their food comes from.

3.6 ECONOMIC IMPORTANCE

Either when the fruit is ripe if it is ready to be sold or until it is ripe if it is to be shipped or processed before marketing, fruit harvesting must be performed. Of course, picking fruit is labor-intensive work that needs great care in handling, as the fruit can easily be harmed. Although fruit such as berries, peaches, pears, and apples are grown and sold on a wide scale to meet the needs of the large networks of supermarkets in the United States, many fruit crops are harvested by the customer. In Illinois, for instance, the strawberry pick-your-own production has risen from 23 percent of the crop in 1967 to 86 percent in 1987. In their yards, several individuals grow fruit trees and berry bushes as well and profit from the fruit they can pick by hand (Satyapal Singh & M.R. Choudhary, 2012).

The ongoing market demand in developed countries for new fruits and vegetables has led to an increase in the amount of fresh produce traded in developing countries. This, in turn, has stimulated the growth of small-scale farms and the introduction of new goods, generating more rural and urban employment, and reducing income inequalities between farms of various sizes. As countries grow wealthier, they are growing their appetite for high-value commodities. In developing countries, the impact of revenue growth on consumption is more pronounced, with higher shares of extra revenue projected to be spent on food items such as meat, fruit, and vegetable items relative to developed countries.

One of the essential and expensive divisions of horticulture is fruit growing. Since then, it has been practiced in India. Art and science have now grown into one of the most professional and intensive ways of land use that are worst. The quality of life of the citizens of nations is determined by the production and per capita conservation of fruits. There are many economic benefits to fruit growing (Satyapal Singh & M.R. Choudhary, 2012).

With high productivity in fruit production comes:

- High net profit: Although the original cost of developing an orchard is high, due to higher efficiency or high product value, it is accounted for by higher net profit. Since height is the original cost of creating an orchard. It is accounted for by higher efficiency or quality products value. Although the average yield per tree is lower (4 to 5 kg / plant) in cashew due to its market value (200 to 250 Rs / kg), higher economic returns are obtained (800 to 1000 Rs / plant).
- High yield per unit area: More production from fruit crops is estimated from a unit area of land than any of the agronomic crops. Grape, banana, and papaya yields are 10 to 15 times higher on average than agronomic crops.
- For various agro-based industries, high productivity fruit farming offers raw materials – preservation and canning (fresh fruits), pharmaceutical industry (Jamun, Papaya, Aonla), coir industries (coconut husk), packaging and transporting industries ,etc.

Figure 3.7: Wages, taxes and services are charged by farm enterprises. They also buy from their local communities' goods and services, multiplying their effect and adding to off-farm jobs.

Source: https://www.farmcrediteast.com/knowledge-exchange/Blog/todays-harvest/economic-impact-of-fruit-and-vegetable-production-in-the-northeast

- Efficient use of resources: Growing fruits is perennial in nature, enabling growers to remain active in farm operations throughout the year and to make full use of resources & assets such as machinery, manpower, land, and water for commercial production during the year relative to agronomic crops.
- Use of barren and wasteland for production: While most fruit crops need perennial irrigation and good soil for development, there are several hardy fruit crops in nature; Jamun, Phalsa, Aonla, Custard apple, Ber, Mango, etc. which are produced on poor undulated and shallow soils considered unfit to produce grain/agronomical crops.
- Foreign exchange: Many fresh fruits, refined goods and spices are shipped to several countries with strong foreign exchange earnings.
- Per Unit Yields are Height: Well-maintained and proven orchards produce better yields than many of the crops in the region. More yield/ revenue than any of the agronomic crops is realised from a unit area of land. The average papaya and banana yields are 10 to 15 times higher than those of agricultural crops. Out of the

overall area under cultivation, grain crops account for 72%, with 210 million tons of output. Although the fruit corps produces over 60 million tons from just 0.7 percent of the total cultivated area due to extremely high productivity.

Chapter 4

Planting Systems and Transplanting of Fruit Trees

CONTENTS

4.1 INTRODUCTION

Orchards are a long-term investment that requires a great deal of experience and preparation. In orchard planting, various planting patterns or planting arrangements may be used. The typical forms are the rectangular and square or hedgerow, diamond, or quincunx, and triangular or hexagonal in row-planted fruit trees. When linked by an imaginary line, such planting arrangements are centered on the geometric shape that the nearest hills create. These are also used in such perennial crops as coconut and oil palm. During the preparation and planting of a new orchard, various aspects such as location and site choice, soil, and subsoil composition, planning of appropriate types and types of fruits, sufficient planting distance and obtaining of plants from reputable nurseries should be given the utmost care and attention.

Site and location

Proper site selection is essential. Based on the following parameters, selection can be made:

- The area must be in an excellent fruit growing area so one might benefit from other growers' expertise and get the advantage of selling the commodity with other fruit growers through cooperative societies.
- A market close to the region should exist.
- The climate must be suitable for growing the fruit crops selected.
- A sufficient supply of water should be accessible all year round.

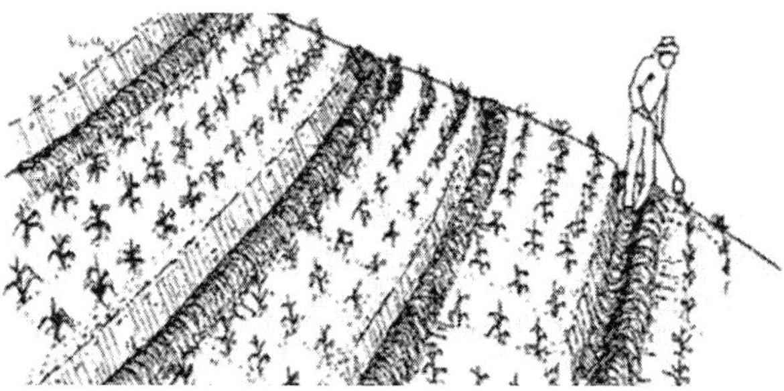

Figure 4.1: The choice of a planting method relies on the climate and weather patterns of the site of the orchard, the species of plants to be cultivated, and the management techniques of the orchards.

Source: http://ecoursesonline.iasri.res.in/mod/page/view.php?id=1578

Even before the grower chooses a place to set up a new orchard, he/she should have examined the following aspects:

- The appropriateness of the soil, its fertility, the composition of the subsoil and its depth.
- During the rainy season, the site should have good drainage and therefore no water stagnation. Water supply must be of excellent quality.
- There must be adequate transport facilities within reach, by either road cr rail.
- If the weather conditions are sufficient for the fruits to be grown and if the site is free of restricting variables such as intense hot winds, cyclones, frosts, and hailstorms.
- At any given time of the year, whether there are seasonal gluts or overproduction.
- If there is guaranteed demand for the fruits to be produced in the market.
- If the orchard is a new venture or if other growers already exist.
- Disposability of labor.

4.2 PLANTING SYSTEM

After evaluating the slope of the property, purpose of using the orchard space, efficiency, etc., the planting method to be implemented is selected. Typically, for fruit trees, six planting systems are suggested. These are divided into two main groups: vertical row planting pattern, and alternate row planting pattern.

4.2.1 Vertical row planting pattern

Square system

It is the most widely used and easy-to-lay method in the region. In this method, the distance from row to row and plant to plant is the same. The plants are at the correct angle to one another, each unit creating a square of four plants. After the orchard is established, this method promotes the inter-culture in two ways. In square planting, the corners of a square that has 4 sides of equal lengths occupy one plant or a group of plants in a common hill. In square planting, a spacing of 10 m x 10 m would lead to a crop area of 10 rows and 10 cross-rows, all being 10m apart and perpendicular to one another. The plants always form rows that are approximately 7.1 meters apart, diagonally. In orchards that use square planting, there is indeed a risk that the canopies of neighboring trees will intersect over time. Light penetration would be hindered as a result. As with mango, the last resort, to expand the lengths of the remaining plants, is thinning, eliminating excess plants. For rectangular planting, this often occurs, which is achieved by eliminating alternate rows and cross-rows.

The resulting new planting pattern would be 20 m x 20 m squared by cutting alternating rows and cross-rows that are spaced 10 meters apart. However, by transforming the planting structure from 10 m x 10 m square to 20 m x 20 m quincunx, complete elimination of alternate rows and cross-rows can be deferred. This is achieved by cutting alternative trees with slanting lines. Therefore, the current nearest plant-to-plant gap (filler to the main hill) will be around 14.1 meters, while there will be 10 meters for both cross-row to cross-row and row-to-row distances. The filler hills are gradually eliminated, leading to a square planting pattern with a spacing of 20 m x 20 m.

Intercultural activities, spraying, harvesting, etc. can be performed quickly and efficiently under this method. It is possible to plant fast growing fruit trees, such as guava, banana, and papaya, during the early life of the orchard. It is possible to grow inter-crops such as coriander, cumin, turmeric, ginger, vegetables, and other spices conveniently, and irrigation can be completed in two directions.

Advantages of this method include:

- It is the simplest and most famous one.
- The plant to plant and row distance is kept identical in this method.
- Plants are precisely right at each other's angle.
- Inter-culture operations can be performed in both ways.
- Adequate inter-cultivation space for remunerative crops such as vegetables.

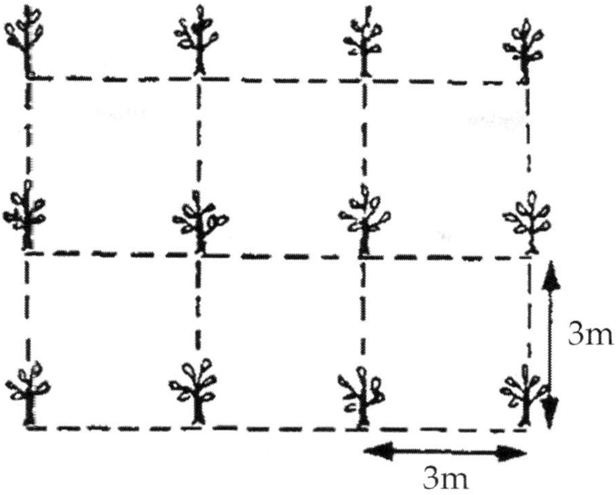

Figure 4.2: The layout's square system is simple to draw and presents each fruit plant with equal space. The space stays unused at the center of the square, which is the primary disadvantage of the scheme.

Source: http://ecoursesonline.iasri.res.in/mod/page/view.php?id=1578

Rectangular system

The plot is divided into rectangles rather than squares in this scheme, and trees are cultivated in straight rows running at right angles at the four corners of the rectangle. This system, like with the square system, also promotes inter-culture in two directions. The only distinction is that more plants in the row can be integrated in this scheme, keeping additional space between the rows. Even though a rectangle has two sets of opposite sides offering different lengths, the rectangular arrangement is identical to a square pattern. A 10 m x 12 m rectangular plantation would imply that two adjacent rows would be 12 meters apart and plants should be 10 m apart in each row; perpendicular to these rows are 10 m spaced cross-rows of plants 12 m apart. The planting pattern is commonly termed a hedgerow when the plant-to-plant distance within the row is 1/3 to 1/2 of the distance within rows.

The advantages of this system include:

- It can be laid out in a rectangle format.
- More row-to-row room.
- Inter-cultural practices may be conducted in both ways.
- Plants get space and sunlight properly.

4.2.2 Alternate row planting pattern

Hexagonal System

The trees are cultivated at the corners of the equilateral triangles in the hexagonal system. In this way, six trees create a hexagon with another tree in its middle. This method, while a little daunting for implementation, incorporates 15 percent more plants. With this method, farming of land between the tree rows is feasible in three directions. In general, this scheme is practiced where the land is expensive and very productive with ample supplies of irrigation water. The planting arrangement's triangular or hexagonal pattern is centered on an equilateral triangle, a triangle with three equivalent sides, which is formed by linking 3 neighboring plants to an imaginary line. A perimeter that is hexagonal in shape will also design six of these triangles that are linked to a specific center hill. There should be rows that are 5 meters apart with 10-meter plant-to-plant gaps and cross-rows that are 8.66m apart perpendicular to these rows. But adjacent plants are spaced 17.32 meters within each row and adjacent plants would be 10 meters apart within each cross-row. The rows with the broadest plant-to-plant distance (17.32 m) are aligned in the east-west direction to optimize light exposure.

In general, this scheme is practiced where the land is expensive and very fertile with ample supplies of irrigation water. While 15 percent more trees can be planted over the square system in a unit area by this process, fruit growers typically do not follow it, as it is difficult to lay out and planting can not be done as effectively as in the square system in the plot.

Advantages of this method include:

- It integrates 15% more crops than the square system.
- At the corner of an equilateral triangle, plants are cultivated.
- There are six trees planted to create a hexagon.
- In the middle, the seventh tree is planted and named a septule.
- This needs land that is fertile.

The main disadvantage is that laying it out is hard and cumbersome.

Quincunx system

This scheme is just like the square system, except in the middle of each square, one extra tree is planted. This system almost doubles the quantity of plants per acre compared with the square system. Fruit trees such as mango, peach, guava, phalsa, kinnow, papaya, etc. can be cultivated in the permanent trees as fillers that provide the grower with an extra revenue in the early life of the orchard. The filler trees are forcibly removed when commercial fruiting starts with the main orchard trees. The shape that is formed is a diamond by connecting 4 nearest hills with imaginary lines, which include 2 main hills and 2 filler hills. With quincunx planting 10 m x 10 m, the length from main-crop to main-crop will be 10 meters while 8.66 meters will be the distance from filler crop to main-crop. The field area would have rows and cross-rows seen from the outer peripheries that are 5 meters apart with alternating filler crop and main crop.

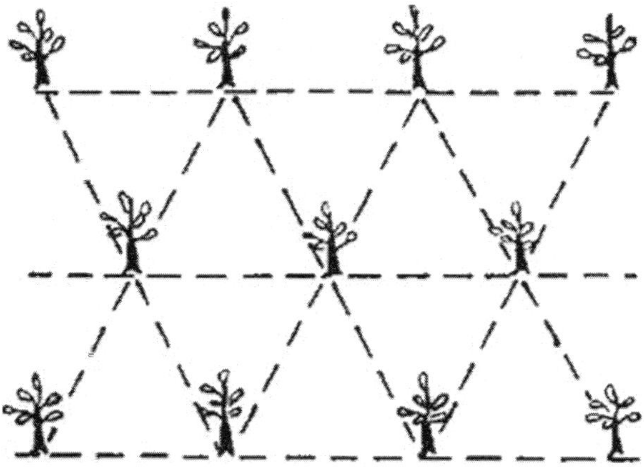

Figure 4.3: The drawbacks of the quincunx planting method are intense competition amongst primary and filler fruit crops for air, sunlight, nutrients, and water, along with overcrowding of trees and overlapping of branches.

Source: http://ecoursesonline.iasri.res.in/mod/page/view.php?id=1578

Contour system

In hilly places with high slopes, this system is commonly practiced, but it is very identical to the rectangular/square system. In such situations, with only a small slope, the trees may be well cultivated in lines surrounding the contour of the soil. Cultivation and irrigation are then only performed across the slope of the land as this process decreases the chance of soil erosion. The construction of this method is carried out as in the square/rectangular system, first by placing the base line at the lowest point and then labeling the trees from the base to the top. Where the slope is greater than 10 percent, bench terraces are utilized.

Triangular system

As in the square method, trees are grown in this system, but the plants are planted in the 2nd, 4th, 6th, and other alternate rows halfway between the 1st, 3rd, 5th, and other alternate rows. This method provides the trees and intercrops with more open space. Other than having more open space for the trees and for intercrops, this system has no advantage over the square system. It is not only a difficult design, but it is also difficult to cultivate in the plots under this method.

4.3 TRANSPLANTING FRUIT TREES

The best and most important plants people can grow in their garden are arguably fruit trees. These trees provide us with excellent nutritious food, wonderful scenery, animal sanctuary, and an excellent excuse to walk around the orchard (or garden). Most fruit trees have common sighting factors, and although there are some variations, trees placed in full sun and cultivated in fertile soil that is well-drained will always do the best. Sighting for the best interest of the tree, however, can interfere with the landscaping plans or hinder other elements of the garden. Therefore, if we truly wish to grow a specific tree and if sacrifices need to be reached, then the way to go is to err based on good growing conditions over all else, so what is the point of planting a fruit tree if it will not perform and flourish?

However, the most diligent thought and preparation often results in planting a fruit tree in the wrong position. It could be in a spot where the farmer just does not want it anymore or in a place that is just not good enough for the tree's optimum health and growth. For whatever cause, problems with fruit trees sometimes become common over time. When this scenario arises, we are simply faced with only two courses of action: leave the tree where it is and comply with the repercussions or transplant the tree. It is costly to purchase an established fruit tree; moving a tree would save money and expanding time rather than cutting it out or abandoning it to die a slow death in a place it hates (Hancock, 2008).

Just because a tree has been continuing to grow in a specific position for a period (possibly years), does not mean that it cannot be removed and transplanted efficiently with a little foresight and concern. It is crucial to recognize; not just surviving and shifting the tree to a great spot can make a world of difference. A fruit tree should flourish. Even if a tree thrives, but the place is inconvenient, it is better to move the tree instead of being continually irritated when you go into the garden. Yeah, it is possible to move the tree and kill it because of the transplant shock, but the cost is worth the effort at the end of the day (Hancock, 2008).

4.3.1 Steps taken to transplant a fruit tree

Establishment of a new location

A yard reconnaissance to find a spot where the transplanted tree can go is the first task that must be done. Initially digging the tree out and then choosing where to go would just cause the tree undue stress and could lead one to make another move that is ill-considered. A location that enhances its current position is the logical target for where the tree ought to go. A hole must be pre-dug in the new location and any necessary excess soil, compost, or mulch must be pre-positioned close to the hole. Timing and winter are usually a good time for transplanting along with area, as most fruit trees during this time are semi or completely dormant. If the tree is fruiting in the winter (such as some citrus in the sub-tropics), and following fruiting, and before the start of the new spring growth, transplanting can be achieved. Thus, at the most critical time of the year, growth will not be "checked" and the following growth season after transplantation will have an immediate impact on getting the tree off to a good start in its new position.

Figure 4.4: We must supply a fruit tree with the best growing environment we can provide to do its job and start providing us with edible, good tasting, healthy goods. Therefore, to promote the best growth environment, often relocating the tree is required.

Source: https://www.selfsufficientme.com/fruit-vegetables/how-to-transplant-a-fruit-tree-in-4-easy-steps

Pruning

Good pruning must be done prior to transplanting to minimize water loss via evaporation and promote a flush of new growth. It is easier to prune when the tree is in its current position since it already has a solid footing, so it is easier to remove branches and prune before transferring means less tree to move and a tidier tree to move. Where possible, the tree requires a particularly good trim cutting back to right above nodes or possibly new growth sites, checking that any crossing branches or dead wood are also pruned and attempting to open the tree center to allow adequate ventilation.

Digging out

Move around the tree to dig down until a complete circle has been cut, then use the same spade input to start shifting around again to start raising and loosening the root ball. Based on the volume of the tree, with the spade or by raising the trunk, it may be easy enough to lift out at this point. For larger trees, to completely loosen the root system and surrounding soil, it might be appropriate to gently force the tree over on one side and back in the other direction. It will minimize transplant shock by preserving as much of the root ball as feasible. Having said that, the tree always must be manageable, because it is not possible to dig out a huge slab of earth with the roots if it is not shifted because of the weight (unless one has equipment like a digger); so, if it is manpower one is using, weight is a definite factor when digging out the tree.

On a fruit tree, the two often significant roots are the feeder roots and the taproot. The taproot is the root that normally lies under the trunk that goes straight down. It can be bad to split the tap root awfully close to the base since it is like a major artery for the plant. The surface roots are feeder roots that look like a nice weaving nest of roots. Most of the water and nutrients required for the tree are collected by such roots. When moving a tree, feeder roots can potentially be broken and pulled away, so caution must be taken to maintain as many feeder roots preserved as possible (Hancock, 2008).

It is time to begin digging with the two significant root systems in mind. For this job, a spade with a sharp edge is ideal as it will be utilized to cleanly cut the root system. This can be done by forcing the maximum depth of the blade securely down with the spade bent slightly inward, not directly down and about 45cm away from the base of the tree. If a root is reached and the spade struggles to cut across, then it can get through with a few hard step-downs on the spade. Ultimately, if it seems like it is still attached to the surface, the tree will feel like it has "come away," then it probably still is thus, some deeper roots are likely to have been lost. Only locate these focal roots and cut them by moving the tree to one side (or another). The tree can then raise out with some effort and have its tap root preserved; most or everything. This should be done with as little damage to the roots as possible when raising the fruit tree (one might need some help). Soil may be lost below the roots of the feeder, but one should try to hold some of the topsoil attached to the roots of the feeder (T. Radha & Lila Mathew, 2007).

Replanting

Additional equipment may be needed for transporting the tree to make the process simpler and mitigate disruption to the root ball, based on how far the tree requires to be transported. When transporting the tree to its new spot, equipment such as a cart, a simple tarp, wheelbarrow, or open growbag with handles, can be helpful. The farmer should always position the central tree with a tarp, and then move the tarp to the transplant site. Of course, if the distance to the place of the transplant is substantial (especially if moving homes, for instance), then the back of a utility/pick-up is ideal.

Before beginning to backfill, one can put the tree in the new hole and take the time to guarantee that the orientation is as they want, as this is the opportunity to rotate the tree so that it is best placed. If the tree is planted on dry soils, then it is also a good idea to pre-wet the area. This is accompanied by the backfilling of the new hole to ensure that the tree trunk is at the correct ground height as before and that the soil does not intrude above the initial collar mark along the trunk base. To enable the tree to cope effectively with transplant shock, one should then gently firm-down the soil and cover with some sufficiently matured compost and water-in using a watering can full (9 liters) of a seaweed solution (recommended dose). To preserve moisture and shield the freshly transplanted roots from the elements, one should mulch well with Lucerne or sugar cane. Staking the tree at this stage may also be beneficial before a stable root structure has been re-established by the tree. With any material or twine from one to another, two stakes driven-in on either side of the root ball (not through it) fits well in a style tie (T. Radha & Lila Mathew, 2007).

Finally, one should periodically water the recently transplanted fruit tree till it is mature, and this is especially important in warmer climates because it can be very dry and warm on some winter days. It is possible to extract some fully dormant fruit trees (including stone fruit) and wash the roots clean to becoming bare rooted before transplantation (deciduous fruit trees are also sold like this), but it is still advisable to transplant with minimal root disturbance. It is prudent to extract fruit at the pruning stage so that the energy of the tree can go into developing new growth and overcome transplant shock rather than producing fruit.

Chapter 5

Orchard Establishment and Management

CONTENTS

5.1 INTRODUCTION

An orchard is a planned estate of trees or bushes that are kept up for food creation. Plantations include organic products or nut-creating trees which are largely developed for business creation. Plantations are likewise here and there an element of huge nurseries, where they serve a tasteful just as a profitable reason. A natural product garden is for the most part inseparable from a plantation, even though it is determined to a more modest non-business scale and may accentuate berry bushes in inclination to organic product trees. Most calm zone plantations are spread out in an ordinary matrix, with a chopped or mown grass or exposed soil base that makes support and organic product assembling simple.

Most orchards are planted for a solitary assortment of fruits. While the significance of presenting biodiversity is perceived in woodland ranches, it would appear to be useful to present some hereditary variety in plantation estates also by planting different trees through the plantation. Hereditary variety in an orchid would give versatility to pests and infections similarly as in timberlands. Orchards are in some cases placed close to waterways where climatic boundaries are directed and bloom time is hindered until the ice threat passes (Louise Ferguson & Elizabeth Elliot Grafton-Cardwell, 2014).

The achievement of an orchard relies generally upon the underlying foundation of plants. A grounded plantation keeps away from numerous issues identified with plant development and execution. Accordingly, one can forestall numerous dangers through suitable plantation foundation practice.

5.2 ORCHARD ESTABLISHMENT

An orchard is a long-term investment that requires keen planning and expertise. During the planning and planting phase, one should concentrate a lot of attention should be given to different aspects for example picking the location and site; nature of the soil and sub-soil; planning of suitable varieties of fruits, establishing proper planning distance, and finally procuring plants from dependable nurseries.

Site selection and preparation

The achievement of an orchard is just tantamount to the arranging and site planning that goes into it. This is a basic proverb, yet one that is frequently neglected by beginner and experienced cultivators the same. Alternate ways and erratic methodologies can bring about not exactly ideal development and different issues during the orchard's life. It is simpler to alter a site before trees are planted than it is once they are in the ground.

Size

Plantations should be enormous enough for packing house groups or work contract-based workers to use teams productively for fundamental activities. To move a team a few miles during a bustling season for a couple of hours' work is normally not conservative; accordingly, pressing houses might be hesitant to contract for dealing with crops from little acreages. Then again, an enormous square of trees of a similar assortment may not be supported, diminished, or reaped inside a brief timeframe fundamental in light of the enormous quantities of talented specialists required. It is generally best to plant no less than 5 sections of land nor more at that point 15 to 20 sections of land of one assortment in one square.

Weather

While choosing a plantation site, think about climate also, microclimate impacts. The site should not be as cold during pre-spring or late-winter when flowers blossom or on the other hand, tender natural products can be harmed by ice. Be that as it may, during the torpid season the site should have a period of chilly climate that fulfills the chilling necessity fundamental for typical organic product set and development in the following developing season. Regularly, winter chilling is estimated by figuring the number of hours at or underneath 45°F (7°C) during the torpid season.

It alludes to the climatic states of a district. Temperature is the main factor that influences the success and the failure of the natural product crop. A diverse organic product needs an alternate scope of temperature. Apple, pear, plum, and peach can be filled effectively in a cool atmosphere (7.5-12.5°C) and these trees shed their leaves in a particular period and produce new leaves and blossoms after a rest period. Mango, banana, papaya, and pineapple require a long-developing season and high temperature (18.5-23°C) without ice and snowfall (Louise Ferguson & Elizabeth Elliot Grafton-Cardwell, 2014).

The reaction of organic product development differs with the change of temperature. Climatic moistness prerequisites vary from plants to plants, for example, banana and papaya require high mugginess however the mango require less dampness. Organic products requiring high stickiness can't be made due to bone-dry zones. High moistness additionally favors bacterial and contagious illnesses. The sum and circulation of precipitation in the district should likewise be considered before the choice to develop the fruit crops. Elevated precipitation and waterlogging conditions are bad for organic product crops. Precipitation at blooming/blooming influences fertilization by washing the stigmatic liquid and dust from the stigma. The time before collecting is additionally viewed as bad for the post reap a life of natural product since it builds the delicate quality (banana and dates) and contamination by organic product flies. If the plantation is planned with a nearby estate, the irregularity of fruit fly infections increases, for instance, guava planted at close distance in moist areas generally stays a survivor of anthracnose disease. In citrus, the issue of organic product drop gets extreme at high dampness. Then again, the vast majority of the natural product crops can't fill appropriately in the locale where rainfalls are low, so need extra water system water to satisfy the plant prerequisites. Hot, high-speed winds and hail-storm are additionally unsafe to organic product plants.

Aspect

Preferably, a plantation should be planted on a south or south-west-bound incline. This angle guarantees the trees will get a lot of daylight consistently and that there will be sufficient seepage. Different perspectives might be thought of being that as it may if the south or south-west slant isn't accessible. Contingent upon different components, for example, introduction and concealing, more solid natural product assortments should be picked to adapt to the somewhat cooler conditions on these inclines.

Exposure

Fruit tree blooming is incredibly defenseless to the breeze blow, especially the previous blooming assortments. In uncovered spots, a windbreak, for example, a hedgerow, can help shield the natural product trees from harsh winds, so you could try at planting one preceding your organic product trees go in (or simultaneously). Another risk to organic product tree bloom is ice. Late ice can obliterate the entirety of the blooms in a plantation; once more, early blooms are especially in danger. To dodge the danger of ice however much as could reasonably be expected, plant later blooming assortments. On the off chance that you have your heart set on early assortments, plant these at the highest point of the slant. Trees at the highest point of the incline will get the most daylight for the longest piece of the day; hence, ice will liquefy rapidly from them.

Drainage

Fruit trees for the most part are affected by waterlogging, so the ideal area for them would be on free-depleting soil. This is halfway why inclines are useful for natural product trees. A decent test to discover how well your ground channels is to make an opening in the ground, around a foot down, and leave it for seven days. If you can see water pooling after this time, the area is most likely excessively wet for solid natural product trees and it very well might merit considering elsewhere. If your waste is alright, however, your soil has high mud content, it could merit diving in some all-around spoiled excrement at every area where you will plant the trees. This will help improve the confined soil structure for the tree while it builds up. Do likewise if your soil is very sandy.

Boundaries

It is additionally worth considering the kinds of limits you have around your plantation site. Fruit trees need a great deal of daylight and won't progress admirably whenever planted near other tall trees, (for example, a fence limit or neighboring forest) since they will be over concealed. Trees filled in shade will have screwy development (as they go after daylight) and might be more vulnerable to certain illnesses (especially parasitic infections) as the microclimate around them will be cooler and wetter.

Tree selection

As much as it merits thinking about your area, it is similarly essential to pick your trees astutely. The accompanying pointers should help you settle on the right choice. Make sure to interact with your tree provider to affirm your decisions, they are there to help.

Rootstock

Fruit trees are provided on various rootstocks. Rootstocks give an anticipated development rate and structure and imply that natural product trees can be planted by and large if the suitable rootstock is chosen.

Rootstocks change from amazingly overshadowing to incredibly enthusiastic, thus the subsequent developed tree will be anticipated tallness; little going to extremely huge. There are favorable circumstances and weaknesses in the utilization of various rootstocks. Predominating ones are anything but difficult to prune and gather from, yet have a helpless root framework meaning they require marking constantly. Vivacious rootstocks can prompt huge inconvenient trees that are hard to collect from and prune, nonetheless, they will create solid roots and will be hardier to antagonistic climate conditions. The rootstock picked will impact how broadly the trees should be planted separated as an impression of how huge the trees will in the long run become.

Varieties

Which types you go for totally rely upon what you need to get out of your plantation - regardless of whether it is for industrial or kitchen use, for instance. The solitary contracting factor with varieties is whether you need to pick later blooming ones if your site is extremely uncovered (see above). At whatever point conceivable, plant probably some conventional, neighborhood varieties in your general vicinity. Conventional assortments are at risk of being lost as more present-day ones are picked in inclination. Your neighborhood provider will have the option to guide you.

Pollination

Many fruit trees are 'self-sterile'; that is they require another assortment to be available for the fertilization to happen so that fruits are delivered. The other alternative is 'self-pollinating trees, for example, those trees which produce organic products with or with no fertilization from another tree. Indeed, even self-pollinating trees profit from some fertilization as this can improve yields.

Fruit trees are organized into fertilization gatherings depending on when they come into bloom. Natural product trees must be pollinated by assortments in a similar gathering as them, or the following one up or down. As such, bunch 2 trees can be pollinated by bunches 1, 2 a, and 3 yet not 4.

Most self-sterile trees are 'diploid'; that is, they just require one other variety for fertilization. A few, in any case, are trickier and are called 'triploid'. Triploid trees need in any event two different varieties in a similar fertilization gathering to guarantee a harvest of product. The fertilization requirements of fruit trees should be on the tree grade.

Land preparation

Poor land readiness and wrong format force repeated issues. Hence, the select fields should be profoundly furrowed and afterward harrowed to uncover the perpetual weeds, roots of perpetual shrubberies, and weighty lumps. On the off chance that the litchi estate must be undertaken on bushy/ minor land, freeing from brambles and backwoods stuff is basic before profound furrowing. Under hillocks furthermore, steep disordered land (30-40% incline), form digging rather than profound furrowing should be done. This activity gives suitable tilth to tender roots for their sound turn of events.

After disturbing, appropriate leveling of land is done and a delicate slant is given one way to encourage the water system just as waste of overabundance water during rain. The soil that has waste issues should be given satisfactory digging from the earliest starting point to maintain a strategic distance from genuine harm to tender plants because of water stagnation. On the slanting area with an inclination up to 10 percent, it is prudent to plant the trees on hills or in channels. This encourages the executive's activities and improves seepage.

Where the land is all the more steeply inclining, patios should be set up, even though this makes furrowing extensively more troublesome. It is likewise fitting to have grass or stone-lined water system trench to forestall soil disintegration. Profound furrowing of the land in dry climate preceding planting will guarantee uniform development, what's more, it is fundamental in recently developed soils where a minimized layer has as a rule been framed. It is attractive, yet not basic to utilize cover crops before planting and proceeding with the equivalent for the initial 3 or 4 years. They will forestall weed development, go about as a windbreak for the youthful trees, and give a wellspring of "in situ" mulching.

Land preparation

- The land should be cleaned properly to allow the free movement of people and machinery.
- The removal of trees, bushes, and creepers is necessary.
- Thorough preparation of the soil in the area designed for growing fruit plants.
- Deep harrowing and plowing on the virgin land.
- Repeated plowing of the land to bring the soil to a fine tilth.

Soil reclamation

In ordinary just as antacid soil, plants don't need liming be that as it may, if the soil examination shows the lower pH than required (5.6-6.0), lime should be applied and if there is magnesium insufficiency, dolomite, should be utilized to improve the soilsoil pH. If the soilsoil is acidic, weighty lime applications might be vital. Two-third of the suggested amount of lime must be dispersed over the planting zone, blended in with the topsoil, and afterward furrowed as profound as conceivable, in any event, 9 to a year before planting.

Calcium (lime) moves gradually downwards into the soil and should thus be applied into the profundity of the root zone. On the off chance that a lighter lime application (2-4 t/ha) is required, the lime can be applied into the soil in any event three months before planting. The lime should possibly be applied when the pH is lower than the objective soil (water) pH. In basic soils with a pH of 7-8, rehashed utilization of sulfur might be made at the pace of 500 kg/ha, until pH falls to 5.6.

Layout plan

The aim of placing a layout or developing a layout plan is to ensure the provision of adequate space for plants; to ensure an adequate number of plants are accommodated; allow easy intercultural operations and determines the system of planting.

- A layout refers to marking the position of the plant in the field.
- Careful preparation of the layout plan in the orchard should be done, it should be based on horticultural expertise.
- The orchard layout plan includes the system of planning provision for orchard paths, roads, water channels, and farm building.
- A sketch of the proposed orchard should be prepared before the actual planting.

Method of layout

- For laying out an orchard, using the square system, a baseline is initially established and this is followed by marking the position of trees along the line laying wooden stakes in the ground.
- Another baseline at a right angle to the first baseline is marked along with the other edge of the field with the help of a carpenter square or a cross-staff.

- A measuring tape can be used to draw the right angle.
- One end of the tape is fixed at a three-meter distance from the corner along the second baseline for a four-meter distance. The diagonal distance between these two points should be five meters.
- The wooden stakes are put in the ground at the desired distance along the second line
- All four rows are thus established and staked. Three men, one putting the peg in the field, and others correcting alignment while moving along the baseline, can easily stake the whole field.

Planting system

As mentioned earlier a layout determines the system of planting. There are various systems of planting generally followed by the basis of the agricultural and climatic conditions and they similarly improve the aesthetic view of the land.

- Square system: This is the most commonly applied method and allows a simple field layout. Plant to plant and row to row distance is the same in this system. The plants are at a right angle to each other, every unit of four plants forming a square. This system allows the intercultural in two directions after planting in the orchard. This type of planting has several advantages. It is the easiest and most common planting system, the plant distance is the same, intercultural operations can be done in any direction and it there is adequate space for inter-cultivation or remunerative crops like vegetables.
- Rectangular system: This system uses rectangles instead of squares and the trees are planted at the four corners of the rectangle in straight rows running at right angles. Interculture operations are also facilitated in two directions similar to the square system. This system unlike the square system is that more plants can be accommodated in the row allowing more space between the rows. The advantages include its rectangular layout, there is more space in between the rows, intercultural operations can be done both ways and the plants are exposed to proper space and sunlight.
- Hexagonal system: In the hexagonal system, tree planting in the corners is in equilateral triangles. Six trees go on to form a hexagon with another tree at its center. This system, though a little complex for execution allows the planting of fifteen percent

trees. Cultivation of land between the tree rows is possible in three directions with the system. This system is generally followed where the land is costly and very fertile with ample provision of irrigation water. The advantages include its accommodation of more than fifteen percent plants than the square system, planting is at the corner of the equilateral triangle, six trees are planted making a hexagon, the seventh tree is planted in the center, and called septuple. This system has one disadvantage it is a difficult and cumbersome layout.

- Quincunx system: It is similar to the square system but there is one extra tree planted in the middle of each square. Plants per acre using this system are double that of the square system. Fruit trees like papaya, guava, peach, and plum can be used as additional trees which are permanent. This provides extra income to the farmer in the early stages of developing the orchard. The extra or filler trees are uprooted when the main orchard trees begin commercial fruiting.
- Contour system: It is a system applied in hilly areas with high slopes but has similarities to the square or rectangular system. In such circumstances, the trees may be well planted in lines that are in line with the contour of the soil with only a slight slope. Irrigation and cultivation are only practiced across slopes since it reduces the likelihood of soil erosion. The system adopts the square or rectangular system. A baseline is first established at the lowest level, and the marking of the trees starts at the baseline moves to the top. Bench terraces are used where the slope is greater than ten percent.

The triangular system has trees planted like in the square system but instead, the plants are in the second, fourth, and sixth, and other alternative rows. The system provides plants in alternate rows are offset half the space between plants in a row. This leads to nine percent more plants than square, and rectangular systems.

Planting

After settling on the entirety of your significant choices on where to plant your trees and which ones to plant, there is then the assignment of really planting them.

Season and tree stockpiling

Fruit trees come in one of two different ways: exposed, established, or containerized. Exposed established trees are accessible in the colder time of year when the trees are inert. These trees should be planted in the cold weather for a very long time between November and March, preferably December. They accompany their foundations enveloped with plastic or something comparative, yet can't be put away like this for any period.

To store uncovered established trees it is important to 'heel' them into a channel in the ground. This includes covering the roots with soil. Containerized trees are accessible lasting through the year and arrived in a pot with manure or soil securing the roots. Since the roots are very much ensured, they can be put away like this in a shielded situation until prepared for planting, which can be whenever of year.

There are inconveniences and focal points to uncovered established or containerized trees. While containerized have the adaptability of having the option to be planted whenever of year, they are twice as costly as exposed established other options. Trees in pots may likewise be root bound, making it harder for them to set up when planted. They likewise require more watering than uncovered established trees, particularly whenever planted throughout the spring and summer.

Stakes and monitors

All organic product trees will require marking for some measure of time after planting. The sort of stake required, and the protection you use to secure the tree will rely on what else you will utilize the plantation space for. If you mean to brush the plantation, substantial stakes and watches should be utilized as most touching organisms (counting deer) will obliterate a plantation of young natural product trees in no time if the trees are not secured. On the off chance that no touching is normal, a bunny gatekeeper and little stake will be adequate while the tree sets up itself.

Planting the trees

The tree ought not to have its underlying foundations uncovered (either from the plastic wrapping or planting pot) until the last possible moment before they go into the opening. Tree roots are harmed promptly after being presented to the air as they dry out so rapidly.

An opening should be burrowed which is adequately profound so the soil level will arrive at a similar point on the tree as where it was at in the nursery (or in the pot). You will have the option to see the old soil line on the tree itself and should be a couple of centimeters underneath the united line (where the rootstock and assortment have been joined).

The stake should be crashed into the ground before the tree goes in with the goal that the tree roots might be orchestrated around the stake, limiting harm to the roots. It should extend in any event 30cm out of the ground, vertically, and ought to lounge around 15cm away from the tree trunk. When the tree has been set in the opening, the soil should be deliberately positioned while the tree is kept vertical. You can shake the tree marginally while the soil is going in to help convey the soil around the roots.

At last, the tree should be approximately fixed with an adaptable bind to the stake, over the united line. The tree should have the option to influence marginally in the breeze as this fortifies the storage compartment.

5.3 ORCHARD MANAGEMENT

The management of trees in an orchard is one of the central parts for getting better returns also, great quality produces. Fruit trees produce natural products paying little mind to human intercession. Natural products house the seeds required for trees to replicate. Nonetheless, for human utilization, it is essential to oversee organic product tree shades to improve the equilibrium between vegetative development and natural product creation, and to keep organic product picking reasonable. An unmanaged tree shelter will deliver fruit products up to a stature of 25–30 feet noticeable all around, which are exceptionally hard to oversee and gather. Shade will assist with building up a solid tree structure that will reduce yield loss while expanding natural product creation and improving natural product quality on the drawn-out premise (Nandwani, 2014).

Pruning

Pruning can be characterized as 'The management of tree structures or fruiting wood or expulsion of any abundance or unfortunate plant parts (shoots, roots, twigs, appendages, and so on) to permit the leftover parts to develop ordinarily or as per the need of pruner'. Pruning is perhaps the most basic methodology for the advancement of a fruitful plantation. Pruning has a lot of gainful impacts as the evacuation of un-beneficial wood, infected, dried, broken, and mismatched. It additionally redirects the progression of photosynthetic aggregates to those plant parts which are fit for bearing organic products. Like preparing, pruning likewise permits more air and light entrance into the inside of the tree covering. It controls the tree size and manages the organic product crop. The pruning time impacts the life of the new tree development, regardless of whether it is energetic or moderate.

First-year trees don't contend well with weeds, which can enormously diminish their development. Weeds go after dampness and supplements and can be disposed of via cautious development, hoeing, or substance splashing. Pre-development herbicides dispose of most, if not all, weeds in spring and late-spring. As the season advances and after a few water systems that will in general drain pre-rise herbicides, weed seedlings may develop to where they need different controls (Nandwani, 2014).

Exceptional consideration should be taken to control perpetual weeds, for example, johnsongrass, bermudagrass, and bindweed, which are hard to dispose of once they become grounded. Now and again screen another plantation for harming pests, for example, peach twig drill, bogus chinch bug, furthermore, vermin, which can genuinely lessen tree development.

Rabbits regularly harm recently planted trees by eating foliage or biting bark off trunks and should be controlled whenever there's any hint of tree injury. Similarly, control gophers if their presence is noted (for the most part by new hills) in the plantation and any enormous fowls (habitually crows) that land in the highest points of youthful trees and break new branches, in this manner distorting trees and easing back development.

Both peach and plum rootstocks frequently send up sucker development from close to or beneath the soil line. These should be taken out while still little when they can be scoured or effectively cut near the storage compartment. Care should be taken not to cause tree bark wounds that can be infected by disease. Light summer pruning once in a while assists with forming healthy trees during the development period.

At the point when they are little, eliminate solid developing shoots situated too low on the storage compartment to later shape into primary side branches or norms. Squeeze back exceptionally upstanding developing shoots, especially in plums, when they are around 8 to 12 inches long. This advances side expanding and tree spreading from the get-go in the life of the tree. Just lively trees should be summer pruned the primary season and afterward just daintily or squeezed back. Weak trees should be permitted to develop and summer pruning should be evaded (Nandwani, 2014).

Training

Training is the positioning of tree appendages in quite a particular manner to deal with the development, as opposed to their expulsion. It can likewise be characterized as 'the advancement of a youthful tree system sufficiently able to bear weighty fruiting with no mischief to branches'. For the advancement of a decent tree system, 2-4 uniformly dispersed fundamental appendages should be chosen to give equilibrium, balance, and more prominent strength. Preparing is profitable as more light and air go into the plant overhang to encourage the appropriate shading and improvement of unrivaled quality of natural products. It likewise shields the trees from wind harm. It is the best hotspot for equivalent dispersion of organic product bearing shoots on the primary appendages of trees. It is likewise useful for ease in social and reaping tasks in the plantation. Preparing evergreen trees is not quite the same as deciduous trees and this is examined below.

a) Evergreen orchards: The term evergreen alludes to "a gathering of plants that hold their foliage during winter". The majority of the evergreens have a solid focal branch pioneer framework, which requires small pruning but to control the plant tallness, increment the stretching thickness, or to cut into uncommon shapes. The suitable identity of development propensities is important before pruning to save the common shape and excellence of any plant. Evergreens trees can be assembled dependent on whether they have whorled branches or irregular expanding designs. New development stretches out from buds that were created from the earlier year's development on the tips of twigs. In any case, a couple of irregular species are equipped for creating new development on both old and new wood branch divides. Prune all evergreens trees before the beginning of new development in the spring or during the semi-lethargic period during mid-summer.

   During the pruning activity, an overall expanding example should be followed to keep up the characteristic state of the tree. Eliminate the dead, infected, or broken branches whenever. Cutting should be attempted in pre-summer or late-spring when new development starts. This permits slices to recuperate and new buds to shape for one year from now. As a rule, specific pruning (one branch at a time) is better when contrasted with shearing/cutting. Shearing makes a formal and mathematical shape that gives a characteristic look yet turns out to be harder to keep up as the plant size increments. Sporadically, an evergreen plant may lose its chief framework. At times another pioneer creates from a torpid (idle) bud, or one of the highest branches will rule and turn into the new pioneer. On the off chance that no pioneer grows normally, tie one of the highest branches straight and train it to turn into the new pioneer. Shorten the encompassed sidelong branches to decrease the internal branch rivalry.

b) Deciduous orchards: Generally, the majority of the deciduous organic product trees are pruned during winter when the plants are lethargic. In winter, the structure of the plant is clearer. It is simpler to discover infected, harmed, and mismatched branches/wood that are discouraging the wind current through the plant shade. These parts should be eliminated as they give a harbor to illnesses and insect pests. Winter pruning will invigorate vegetative development. During the developing period, in great seasons, the tree stores starch in its trunk and roots to permit the plant endurance during winter without the presence of leaves. Such trees will have enough put away food to keep the late spring trees alive.

   If there is weighty pruning in the colder time of year, there is a flood of put away food and profuse development will happen in the accompanying spring. This can be important to re-animate and produce new wood on old or more fragile plants, however, ought not to be rehearsed if you want to restrict the tree's size. After winter pruning, the plant develops all the more energetically, creating vegetative wood as opposed to blooming and fruiting trees. The atmosphere has a huge job at the hour of winter pruning. Young trees are more powerless to harm in hard winters. Thus, pruning ought not to be drilled in pre-spring, late-winter, or after the risk of ice has passed to limit harm in ice inclined regions.

Fruit thinning

Many fruit tree species bear a lot of blossoms which produce unnecessary fruits that the tree can't uphold. Such a large number of organic products per tree can bring about little organic product size and low quality, breakage of branches, the fruit tree stores, and decrease in cool toughness. In some organic product crops, weighty fruiting can mostly or block the blossom bud inception. The cultivators have perceived these as serious issues for natural product creation and tree life span and received fruit diminishing practices as a rule by hand. The term fruit diminishing is characterized as "the evacuation of a little bit of fruit crop from the tree before its full development to build the size and nature of the excess organic product".

Because of natural product diminishing, more starches and photosynthetic aggregate in the excess organic products, thus improving the last natural product size, shading, eating, and keeping quality, and eventually builds the attractiveness of the product. Organic product diminishing during on-year is likewise useful to beat the biennial direction of organic product trees, for example, Mango furthermore, citrus. Also, natural product diminishing additionally lessens the appendage breakage of the bearing tree. Organic product diminishing should be possible utilizing anybody or the blend of the accompanying techniques:

a) Manual thinning: Manual thinning is the manual expulsion of organic product to manage crop load furthermore, increment the natural product size before reap. It is a simple, exact, and safer technique yet, this requires more work for diminishing. It is entirely practical financially with critical expansion in the benefit. Higher worth assortments, for example, mandarin and oranges, can accomplish the financial increase from hand diminishing. Manual thinning should be possible with the assistance of an iron edge having the square with four legs of 0.125 m3 long on each side. In citrus, feeble, flawed, and unfortunate natural products are taken out, leaving 9-10 organic products on each edge. Manual thinning is typically done after the normal organic product drop as ahead of schedule as could reasonably be expected because it shows great outcomes. Hand diminishing in citrus organic products generally begins in late June and early July when the natural product size is almost 30-40 mm in breadth. It expands the net return by 20-40%.
b) Mechanical thinning: Mechanical thinning is an imaginative methodology for diminishing organic product crops prepared to any sort of shaft formed trees, paying little mind to species, cultivar, temperature, and tree age. This innovation is helpful to save the hour of work and is more practical. Mechanical thinning, alone or in blend with one or the other synthetic or manual diminishing, improves the natural product quality, especially natural product size (by improving the source-sink relationship and augmenting the photosynthates parceled to the leftover leafy foods) (by permitting more light admittance to the tree inward shade), sugar (taste), and once in a while immovability for better storability.

The gadget saves work costs, in any case, needed for hand-diminishing. Mechanical thinners incorporate trunk shakers to eliminate the blossoms or green foods drum shakers that have vibrating poles embedded into the plant shade to eliminate the creation of organic products. Mechanical diminishing innovation is minimal effort, requires little expertise, and causes less harm to residuals. Nonetheless, it has a few downsides such as, the expulsion of some best natural products in the stand and leaves a portion of the helpless ones, which are more powerless to wind and frost.

Nutrition

There are seventeen fundamental supplements that are needed for plant development and improvement: carbon, hydrogen, oxygen, nitrogen, phosphorus, potassium, calcium, magnesium, sulfur, iron, manganese, zinc, copper, boron, molybdenum, chlorine, and nickel. Of these seventeen components, all aside from C, H, and O are gotten from the soil. At the point when the soil can't supply the degree of supplements needed for satisfactory development, supplemental compost applications become fundamental. The wholesome necessities of organic product crops are unique in relation to those of annuals. In enduring yields, there is a need to supply supplements for current organic product creation just as for the vegetative parts, which persevere for quite a while.

Fruit trees expect supplements to live and flourish. At the point when at least one of these supplements is lacking in the soil, the tree won't arrive at its full scene potential, will be more defenseless to infections and creepy crawly bugs, and will have a more limited life expectancy than a tree that has gotten appropriate treatment. Plant supplements can impact the blooming, organic product setting, organic product size, the measure of vegetative development, and other plant qualities. The supplements needed by all plants, including trees, can be partitioned into two gatherings: macronutrients and micronutrients, given the amount fundamental for development and advancement. Macronutrients are needed by plants in bigger amounts when contrasted with micronutrients. The nourishment is essential to give good conditions for the improvement of solid trees that can create more significant returns with better natural product quality (Md Asaduzzaman & Toshiki Asao, 2019).

Fruit trees are not extensive yet have a spreading root framework when contrasted with other field crops. In this manner, the supplement prerequisite of organic product trees is significantly more unique in relation to different yields. Countless elements like tree age, sort of rootstock, and scion assortments make it exceptionally hard to define the specific sustenance program which would apply to every different condition.

Cultivators should figure out how to deal with these supplements for ideal plant development and creation and to limit the unfriendly natural impacts. The most significant is the equilibrium of supplements, when one component is lacking, its nonappearance can adversely influence the plant instruments and consequently block the ideal take-up, usage, or conveyance of different minerals. Then again, an abundance of any component might be harmful to trees, in this manner influencing the accessibility of different supplements in the soil.

Plant roots assimilate the greater part of their supplements from the soil arrangement in an ionic structure. Plant roots assimilate bigger atoms, however, their pace of ingestion is moderate. Hence, manure, by the same token natural or inorganic should initially be separated into its least complex structures that plants can use all the more proficiently (Md Asaduzzaman & Toshiki Asao, 2019).

Besides, appropriate nourishment is one of the most basic variables influencing the foreseen profitability of the cutting edge plantations. In the current situation, supplement inadequacy in the soil has now become a predominant wonder, which signs the diminished yield and nature of organic products. Supplement lack might be ascribed to calcareousness, low natural issue, and the high soil pH. Soil and climatic conditions are helpful to foresee the supplement insufficiency; yet, are not satisfactory to make critical suggestions. Soil and plant examination, for investigating supplement insufficiencies, assume key roles in the fruitful supplement the executives of the plantations. Soil factors are a significant trademark that shows the intermittent spatial fluctuation in orchards.

Irrigation

The artificial utilization of water to the land or soil is known as a water system. It is utilized in developing yields/plantations, support of scenes, and re-vegetation of upset soils in dry territories during times of helpless precipitation. Additionally, the water system moreover has some different uses in harvest creation, for example shielding plants from ice and forestalls soil smallness. To upgrade the water system plan, the factors of climate, water accessibility, and plant pressure should be thought of. These factors may change all through the water system season, expanding the unpredictability in deciding when the water system ought to happen and the measure of water to apply. To decide how much water to apply, the water system needs of the plant must be determined. This sum is climate-driven and is impacted by the stickiness, temperature, and day length time. The interest in water by the tree is equivalent to the pace of evapotranspiration (ET). ET represents the deficiency of water through the soil surface, loss of water through leaves, and stomata that permit the gas trade needed for photosynthesis. In some regions, all plantations have no admittance to the adequate water system, however, satisfactory watering during the early years can help in tree foundation. The time and volume of water applied to fluctuate with tree size, soil, climate, and season. The greatest water use in year four in summer would be 160 liters per tree. Flood plantations a few times each week in sands and one to two times each week in substantial soil is suggested for acceptable yield creation. Diverse water system frameworks can be utilized in plantations as per the availability of capital and volume of water for the water system. These water system frameworks are examined underneath (Zhang, 2017).

Weed management

Weeds are plants developing where they are not needed, and as a rule, meddle with the creation of natural product crops. These weeds contend with trees for water, supplements, daylight, and go about as characteristic living space of pests and diseases. In plantations, the greater part of the weeds is perpetual, having underground roots or stems, which produce the wealth of seeds and stay feasible in the soil for extremely significant stretches. Weeds mostly dispersed through the wind, water, air, and by means of creature composts and so forth. Annihilation of weeds from the plantation is vital as these rival the water and supplements in organic products trees. The presence of weeds in plantations normally diminishes natural product creation. Weeds can be controlled in numerous ways, for example, precisely, seriously, naturally, and artificially. Notwithstanding, a blend of at least one of such techniques gives more compelling outcomes when contrasted with any single one to kill the weeds.

- Mulching: Mulching is a soil administration method in which various kinds of crude materials, like feed, straw, grass, dry leaves, wood shavings, weed scrapings, paper, or polyethylene, are set on the soil surface between the columns of organic product plants. The utilization of mulches likewise influences the soil natural issue content, movement of soil conceived microorganisms, accessibility of supplements, soil compaction, and disintegration. Mulches improve the water holding limit of the soil. The principle reason for mulching is to save the soil dampness, temperature guideline, weeds control, and creation of great natural products. In addition, weeds are controlled without upsetting the soil surface, which limits the water loss by vanishing. Notwithstanding, natural mulches are vulnerable to fire and encourage the hibernation of rodents. Polythene mulching is additionally advantageous to control insects in certain harvests, for example, the inflorescence midge in mango.

- Zero tillage: For the most part, zero culturing is a homestead based activity that decreases the expenses while keeping up the yield and securing the climate. Zero plowing or no-culturing cultivating is a method of raising harvests with no mechanical or manual aggravation of the soil surface. In this framework, weeds are killed by utilizing herbicides, plantation soil is left exposed without development, and subsequently, there is no water infiltration under the soil surface and no development of weeds. This framework diminishes the expense of development and reaping activities. The selection of zero culturing innovation has preferences, for example, decrease in the planting cost, expanded compost, and water use effectiveness (WUE), decreased soil disintegration, and ease in homestead tasks. The most remarkable advantage of zero culturing is the improvement in the natural richness of the soil. Notwithstanding, this framework has a few disadvantages, for example, brisk soil dryness, also, more regular water systems are required.
- Clean cultivation: Spotless development is the activity in which plantation land is constantly kept free from any sort of grasses or vegetation. This framework is useful to control the weeds and improves the soil dampness maintenance; soil air circulation makes the soil aerated by expanding the supplement accessibility. Ceaseless utilization of clean development without expansion of natural issue or any sort of fruit, yield brings about the consumption of soil supplements. Regular clean development is exceptionally gainful when rehearsed inside the period of a solitary developing season. Clean development for broad periods quickens the soil disintegration, loss of natural issue, and crumbling of soil actual properties.

Chapter 6

Use of Growth Regulators in Fruit Production

CONTENTS

6.1 INTRODUCTION

Plant growth regulators (PGRs) are chemicals used to alter plant growth, e.g. increase branching, suppress shoot growth, increase yields, eliminate excess fruit, or modify fruit maturity. Many of these aspects affect the effectiveness of PGR, including how many trees, tree strength and age, dose, length, cultivar and weather conditions consume the chemistry before, during and after application. Plant bio regulators regulate the development and growth cycles of plants, both naturally and synthetically. Plant growth regulators refer to non-nutrient organic compounds that promote, impede or otherwise modify some limited number of physiological processes in plants. Plant hormones refer to plant-release bio-regulators that usually travel from the site of growth to the site of action within the plant and regulate the physiological behavior of the plant at very low concentrations. Five forms of plant growth regulators can be categorized as auxin-related compounds, gibberellins, and Gibberellin Biosynthesis inhibitors, cytokinins, abscisic acid, and ethylene-related compounds. Products that hinder the biosynthesis of plant hormones are also accessible (Apogee, Retain).

Five types of fruit crop growth regulators:

Auxins: These are compounds that facilitate development but may inhibit the growth of high concentrations of lateral buds. Auxins can also be herbicides, in addition to being used as plant growth regulators (2, 4-D ,etc.). The NAA is a synthetic auxin in apple production that can be used to dilute the fruit in order to keep the fruit from dropping shortly before harvesting. Napthaleneacetic acid (NAA) (See Ontario.ca/apples and Tree Fruit Thinning for more detail about the use of diluent ingredients).

Gibberellins (GA) promote cell length, shoot growth, and regulate dormancy. Gibberellins (gibberellins) Promalin/Perlan (containing GA4+7 and 6-benzyladenine) has been used to improve fruit size and resist rust in apples. For the slow maturation of sweet cherries, the ProGibb 40SB and the Falgro-containing GA3 are used to improve fruit firmness and increase harvest time. Gibberellins are used to monitor the flowering of tarts to avoid overproduction. Apogee or Kudos 27.5 WDG® prevents the biosynthesis of Gibberellin (Prohexidione-calcium). Apogee or Kudos 27,5 WDG was used to adjust the morphology and to track the development of strawberries in tree runners (apple and cherries) (Basra, 2000).

Cytokinins: Cytokinins promote different cells. Cytokinins help initiate the branching and activation of buds. It is used as a thinner of apple fruit (Maxcel or Cilis Plus 6-BA). For more information about the use of these finishing products, link to Ontario.ca/apples and find Tree Fruit Thinning. Plant growth regulators (PGRs) are chemicals used to alter plant growth, e.g. increase branching, suppress shoot growth, increase yields, eliminate excess fruit, or modify fruit maturity. Many of these aspects affect the effectiveness of PGR, including how many trees, tree strength and age, dose, length, cultivar and weather conditions consume the chemistry before, during and after application.

Ethylene: Ethylene facilitates the abscission of leaves and fruit, prevents the lengthening of shots and inhibits lateral bud formation. Ethylene is used to make the fruit ripe in apples and cherries. Ethrel® (Ethrel®) is a synthetic compound formed by ethylene. Retaining affects ethylene biosynthesis, allowing the fruit to live longer on trees and to increase the shelf life.

Absicisic acid: Abscisic acid controls the dormancy of the buds and seeds, prevents the development of the shoots and controls the depletion of plant water (C.M. Karssen, L.C. van Loon, & D. Vreugdenhil, 2012).

6.2 USES OF GROWTH REGULATORS IN FRUIT PROPAGATION

The use of auxin-like compounds to promote adventitious radical production in hardwood and softwood cutting is one of the earliest and most important practical applications for plant growth regulators. For many years now, the use of cuttings containing indole butyric acid, alone or combined with alpha-naptha-acetic acid, has been standard practice for both experienced plant propagators and home gardeners. In addition, synthetic cytokines and auxins are an important part of the processing media for the growth of axillary flowers and for the formation of adventurous roots in the growth of fruit plants using aseptic methods. Plant growth regulators are also critical components of in vitro plant regenerative organogenesis and somatic embryogenesis, and are necessary for the application of biotechnology to fruit crop genetic development.

Figure 6.1: Growth regulators are used in the propagation of citrus fruits

Source: https://edis.ifas.ufl.edu/hs1309

The general public understands the importance of growth regulators for plant propagation, but the use of growth regulators for fruit production has been difficult. The protection of pesticides and chemical mechanization in combination with residues, fresh goods sanitation, and general wellbeing of the public and environmental losses impact consumers and advocacy groups. This issue is illustrated in a variety of countries by increasingly stringent prohibitions on the use of chemical materials in agriculture and horticulture. Consumer preferences have since changed. For example, California's Fresh Fruit Advertisements suggest that growth regulators have not been used in crop production. Some also concluded from these social pressures that chemical controls in crop production, including growth regulators, can be greatly reduced, if not phased out, and that greater emphasis can now be placed on fruit breeding and the improved use of genetic approaches in crop protection and management. While the use of growth regulators in orchards or vineyards could be limited in the future, these compounds may still be invaluable methods for plant propagation and genetic enhancement of fruit crops using both conventional and biotech approaches. Its views are expressed in the recent studies on the spread of fruit cultivation and (ii) the early blooming of young seedlings (C.M. Karssen, L.C. van Loon, & D. Vreugdenhil, 2012).

6.3 USE OF PLANT REGULATORS IN ROOTING OF CUTTINGS

Powerful biotic factors regulate the growth of plant hormones in the root. Seven kinds of plant hormones, auxin and gibberellin (GA), are successful shoot growth accelerators, but they are not always root growth accelerators. The new regulatory mechanism hypothesis illustrates the orthodox approach to regulating the radical rise of auxins and gibberellin. Due to its inversely proportional growth rate, the endogenous concentration of indole-3 acetic acid (IAA) is essential to understanding the development of root auxins. Since major IAA is transported from shoots to roots through polar transport, the penetration rate of IAA primarily controls the amount of IAA root cells. The classic IAA root transport mechanism is enabled by the latest discoveries of IAA carrier proteins including AUX1, PIN and MDR. The role of the plasma membrane-located H+ ATPase and its IAA regulation were also determined in the acid growth phenomena due to acidification of the root cell walls.

Figure 6.2: Growth regulators are used in the planting of gooseberries

Source: https://www.mdpi.com/2073-4395/10/4/597

The roots have less unusual GA functions than the auxins. However, GA also plays a valuable role in the natural growth of roots, as artificial GA depletions cause root elongation to be erratically extended and suppressed. Using chemical inhibitors and GA biosynthesis mutants, the requirement for normal root growth of GA is disclosed. The GA function of long and slender root morphology results in the formation of cortical microtubules, cellulose microfibrils and unknown additional factors (s). Cross-conversations between plant hormones have recently been established in signal transduction pathways, primarily in the aerial organ. GA and IAA de-repressing gene expression by degrading protein-repressing genes through ubiquitin-mediated proteasome pathways. The enhancement of the GA1 mechanism by IAA is another interaction between IAA and GA in growth regulation. When final steps in the control of biochemical growth are taken in cell walls, the development and modification of cell walls would be vulnerable to future cross-conversations (C.M. Karssen, L.C. van Loon, & D. Vreugdenhil, 2012).

6.4 USE OF GROWTH REGULATORS IN TISSUE CULTURE

As far as plant cell development is concerned, the main argument is to use special media manipulations to direct plant cell growth in culture due to this plasticity and totipotency. Plant growth regulators are important medium components in the assessment of the mechanism of plant cell development. Plant hormones or their synthetic analogs are the most commonly known crop growth regulators. Plant-raising regulators form Plant-growing regulators. Five primary types of plant growth regulators are used for plant cell cultivation: auxin; cytokinins; gibberellin. Every class of plant growth regulators will be examined briefly. IAA is the most effective natural aid, but its use in plant cell culture media is limited because it is unsafe for heat and light. Indole-3 acid is the most essential naturally occurring auxin. Auxins activate cell division and cell development.

Figure 6.3: Growth regulators have enabled fruit farmers to ensure a continuous supply of fruits

Source: https://www.researchgate.net/figure/Growth-regulators-GA-3-application-stimulates-color-development-and-ripening-of-wax_fig7_313055349

More soluble IAA amino acid conjugates are also used for IAA-related problems (such as Indole-acetyl-L-alanine and Indole-acetyl-L-glycine).

However, stable chemical IAA analogs are more widely used as an auxin supply in plant cell culture media. 2,4-Dichlorophenoxyacetic acid (2,4-D) is the most commonly used auxin and is very potent in most situations. Some auxins can be more powerful or "powerful" than 2,4-D in some instances (Davies, 2013).

There are a number of growth regulators used in tissue cultures. They include: Dicamba 2-methoxy-3,6-dichlorophenetic acid, IAA (Indole 3-acetylic acid), IBA (Indole 3-butyric acid), Picloram (4-amino 2,5,6-trichloropicolinic acid), Picloram (4-amino 2,5,6-trichloropicolinic acid). A replaced phenyloram 4-amino 2,5,6-trichloropicolinic acid Replacement pheny-phenylacetic acid. A substituted phenylate, Cytokinins allow a chemical name distinguishing between cells. Cytokines are usually found in a broad group of structurally similar compounds (purine derivatives). Of the naturally occurring cytokines, two are used in plant tissue culture media. Since they are expensive and relatively fragile (especially zeatin), they are not widely used. There is also increased use of synthetic analogs (Bazylaminopurine), kinetin and BAP. Non-purine chemical compounds, such as well-substantiated phenylureas, are also used as cytokines in plant cell culture media. These substituted phenylureas can also replace auxin in certain cultivation processes. Abbreviation/Name A chemical name Zealinb 4-hydroxy-3-methyl-trans-2-butenylinopurine Synthetic analogs. A cytokininine substituted phenylurea type Chemical Name BAP 6-benzylaminopurines b[N6-(2-isopentyl)adenines] Kinetina 6-furfurylaminopurines Thidiazuron1-phenyl-3- (11,2,3-thiadiazol-5-yl) (Davies, 2013).

Many structurally similar, naturally occurring compounds such as gibberellin are known as gibberellin. They regulate cell elongation and are agronomically important in determining plant height and fruit set. Just a few gibberellins, most commonly GA3, are used in plant tissue culture media. ABA is stopping the division of the cells. Ethylene is a gaseous, naturally occurring, plant-widening regulator that mainly regulates fruit ripening in the atmosphere and is not commonly used in plant tissue production and is used to facilitate distinct developmental processes such as somatic embryogenesis. The greatest barrier to the cultivation of plant tissues, however, is the atom. Any culture in plant cells produces ethylene, which, if developed enough, will prevent the seed from growing and developing. The design and means of closing of the cultivated vessel affect the exchange of gas between the cultivated vessel and the external atmosphere and hence the amount of ethylene contained in the grain. Generalizations of plant-growth regulators and their use of cell-growth media have been developed from early experiments of the 1950s in plant-growth regulators and tissue culture. However, the results of plant growth regulation systems are difficult to predict: this is due to substantial variations in crop reactions between crops, cultivars and even plants are grown under different conditions in the same cultivar. However, such concepts remain valid and have been the foundation for most tissue-growing regimes.

Auxins and cytokinins are the most commonly used plant growth regulators in tissue culture and are typically paired with a connection between the aid and the cytokine that defines the type of cultivation that has been formed or regenerated. Typically, the high ratio of auxin to cytokine promotes root forming, while the high ratio of cytokinin to auxin promotes root formation. The middle ratio encourages the development of the callus. A poor ratio of auxin-cytokinin (shoot formation); High ratio of auxin-cytokinin (root formation) (intermediate ratio of callus formation) (root formation). Effects on the development and morphogenesis of the callus depend on the various ratios between Auxin and Cytokinin. High cytokine auxin ratios encourage root growth, low shooting ratios encourage development. Medium ratios cause the callus to expand slowly without distinction (Davies, 2013).

Forms in culture

Usually, the cultivation of sterile portions of the whole plant is introduced. These sections may consist of parts of the organ, such as leaves or roots, or may consist of particular types of cells, such as pollen or endosperm. These sections are referred to as "explants." Many of the characteristics of the explant are considered to have an influence on the quality of culture initiation. In general, tissue (or tissue at an early stage of development) has grown faster and younger. More details can now be given on several distinct crop varieties most widely used in plant transformation studies. Callus explants can cause an unorganized, rising and dividing cell mass when grown on an adequate medium, typically with both Auxin and Cytokine. If the requirements are right, some plant tissue can be used as an explant. In culture, this distribution can be retained more or less indefinitely if the callus is frequently sub-cultured. During the callus development, there is a degree of differentiation in the type of both anatomy (usually unspecialized parenchymal cells) and metabolism (i.e. modifications which occur during growth and specialization, are, to some extent, revert). The key effect of the differentiation is the lack of photosynthesis in most species of plants. The metabolic profile of the donor plant is unlikely to fit because it has significant reaming effects on the culture of callus tissue. In addition to the usual mineral nutrients, this also includes the incorporation of other ingredients – vitamins and, most notably, the carbon source – into the culture medium.

Callus culture occurs primarily in the dark (lack of photosynthetic potential is not an inconvenience) because light promotes calling for separation. A culture can lose the need for auxin and/or cytokine during long-term culture. In certain plants' callus cultures, this method known as 'habitual' was popular (e.g. sugar beet) (such as sugar beet). Callus crops are also important in the field of plant biotechnology. The handling of the medium auxin-cytokinin ratio can contribute to the development of shoots, roots or somatic embryos that can be used to produce whole plants. Callus crops can also be used to induce cell suspension in plant transformation studies in a number of ways. Cell suspension cultures. The cultures of Callus, generally known as lightweight or friable, fall into one of two groups. The cells are firmly aggregated in a rigid callus, but in a friable one, the cells are just tightly bound together and the callus is quickly and softly torn apart. The friable callus covers the inoculum to form a culture in the cell suspension. Explants of a specific flora or cell type appear not to develop friable calluses, making it impossible to initiate cell suspension (Karssen, van Loon, & Vreugdenhil, 2012).

Often a shift in medium components or a repeat subculture can increase the friability of the callus. Calvary can also be strengthened by growing in a semi-solid (medium with a low concentration of gelling agent) medium (medium with a low concentration of gelling agent). As the cold callus is introduced into a liquid solution (usually the same composition as the solid medium used to develop the callus), it is released into the agitated single cell and cell cluster media. These released cells continue to spacious and differentiate under the right conditions, gradually creating a community of cell suspension. A very large inoculum can be used to start cell suspension in such a manner that the amount of cells released is growing. The inoculum should not be too high, though, as toxic compounds released from weakened or stressed cells may exceed lethal levels. During the subculture of cell suspension, large cell clusters can be isolated. As batch cultures can be preserved reasonably effectively in conical flasks. They continue to evolve into a modern medium by the repetition of subcultures. This leads to the dilution of the suspension and the start of another batch forming period. The sum of dilution can be calculated on an analytical basis for each society. Very long lag or, in serious situations, the death of the moved cells will lead to a great deal of dilution. After subculture, the cells are separated and the biomass of the crop grows in a natural manner before the nutrients in the medium are depleted and/or the poisonous by-products rise to a degree of inhibition, which is called a 'stationary process.' The cells are then moved to the stationary phase. The parameters of the lot growth cycle are therefore essential for any cell-suspension culture (Karssen, van Loon, & Vreugdenhil, 2012).

Protoplasts of Protoplasts

Plant cells are uppermost cells, separated by a cell membrane. Although other sources may be used for this protoplasty, they are most commonly isolated from leaf mesophyll cells or cell suspensions. The cell wall can be separated using two general methods (mechanical or enzyme insulation) (a challenging process without destroying the protoplast). Mechanical isolation leads, while possible, to low yields, poorer quality and poor seed productivity due to cell contamination. Insulation of enzymes in a transparent salt solution plus decaying enzymes in the cell wall is normally achieved with high osmotic content. A combination of cellulose and pectinase enzyme is commonly used, which must be high in strength and purity. Protoplasts are fragile and quickly broken and must thus be carefully cultivated. The osmotic potential is high at least in the initial level. The fluid medium shall not be stirred.

The callus on the support medium is mounted and the mold protoplasts are mounted. This callus will regenerate entire plants through organogenesis or somatic embryogenesis. Protoplasts are ideal candidates for a variety of transformation processes. Root crops from both primary and secondary explants can be produced *in vitro* and cultivated in relatively simple media. Root formation *in vitro* can be boundless, because the roots are eternal. While root crops have been one of the first achievements in modern tissue farming, they have not always been used in transforming plants. Cluster shooting with axillary or daring bud can be grown in vitro with shooting tips and meristem cultivation (including apical meristem shoots). Shooting meristems can be created as a replacement for more widely used low-genotype and more efficient cereal regeneration strategies (seedlings can be used as donors, in particular) (seedlings) (seedlings can be used as donor material). Embryo communities Embryos may be used as explants to develop colonies or somatic embryos. Explants can be used both as seeds and as adult embryos. The most common way to renew monocot plants is through immature, embryo-derived callus. The use of pollen or other agents as an explant can result in the production of haploid tissues *in vitro*. Microspore culture Pollen is discovered in the male gametophyte 'micro-spore.' Pollen may be developed for both calluses and embryos. Two main approaches can be used to create haploid tissue colonies *in vitro*. The first invention involves the use of another as an insightful drug. Another tissue should be grown, a somatic material that protects and maintains pollen (agar is not used to solidify the stock because it contains inhibitory substances) (agar should not be used to solidify the medium as it contains inhibitory substances). Dehiscence of mature anthers is accompanied by the development of pollen-derived embryos. At the right time and under the right social circumstances, the dehiscence of the other depends on its isolation. In certain species, cutting the wall of the anther resolves reliance on natural dehiscence, but of course, it takes a significant amount of time. Another medium can also be cultivated and other pollen may produce embryos, but the productivity of the regeneration of the plant remains very low. Since this process takes time, immature pollen may also be obtained directly from further growth. For any strategy, there are their own profits and challenges (Basra, 2000).

Any positive results have been identified by the use of anthers as explants. However, it is not possible for embryos formed from other crops to come from microsporeal haploid cells, but from somatic anther tissue. While isolated pollen does not allow the development of mixed embryos, productivity and time consumption are reduced. The condition of the donor plant and the time of insulation are essential for the cultivation of the spore. Pretreatment, such as cold therapy, has further strengthened the findings. Pretreatment may be used for the processing or cultivation of anther animals. Based on the fact that the plant growth regulator must be added to the pollen seed medium, the most important organic supplements, such as amino acids, can be classified into two groups. For several cereals with pollen/antiquity regulators, an additional medium is required (rice, wheat, barley, and maize). Microspore explant regeneration can be done by active embryo genetics or by potential embryogenesis callus. Female gametophytes can also begin the cultivation of haploid tissues (the ovules). (This is the egg). (This is the egg, man). It is more active than in other situations of pollen or anthers. No haploid-based plants can be cultivated. This is due to the duplication of chromosomes during the social period. Chromosome replication can be beneficial, but haploid plants are often not appropriate for haploid tissue regeneration since they are often caused by chemical reactions such as colchicine. These plants are sometimes called "di-haploids," since they function in two versions of the genome (Basra, 2000).

6.5 USE OF GROWTH REGULATORS IN SEED DORMANCY

One of the most important adaptation mechanisms in plants is crop dormancy, which prevents seed from premature germination in the face of unfavorable conditions of continuous production. Seed dormancy is controlled by many environmental and molecular attributes. The sun, temperature and water supply depend on the control of seed dormancy or release. The basic response of seeds to environmental stimulation is mediated by a variety of phytohormonal pathways. ABA is thought to control the recruitment and retention of dormant seeds as key phytohormones. Receptive components ABA and Ka ensure that cross-conversations between GA and ABA pathways are formed and that environmental seed is adequately handled. Phytohormonal dormancy regulation mechanisms are similar in dicot and monocot plants. Recently, certain phytohormones, such as auxin, jasmonate, brassinosteroids and ethylene, have also been suggested to control dormancy in plants. Auxin regulators improve ABA behavior and have a beneficial impact on seed dormancy. However, the degree of sperm dormancy is decreased by jasmone, brassinosteroids and ethylene. Recent progress has been made in understanding the dynamic dormancy of seeds regulated by multiple phytohormonal pathways and their components. Studies on germination of seed-sleeping crop species can help ensure the most efficient germination period (Basra, 2000).

6.6 CONTROLLING FRUIT VIGOUR

Fruit growers and salespeople are committed to supplying consumers with stunning, tasty and cheap fruit and to fulfilling their sensory, psychological and dietary needs. Gradually, their designers, components and procedures will be subject to strict specifications on the part of consumers, suppliers and importers.

Growers discover the comparative benefit of farm management and effective management in order to increase yields and fruit sustainability. Its procedures authorize the use of pesticides for plant surveillance purposes. PGR fruit crops such as NAA, GA3, BAP and CPPU have been used primarily to improve fruit properties such as size and strength, improved shape and color, reduced seed count, improved yields, reduced volatility and prevented stock disorders. Other PGRs, such as PP333, maintenance and cyanamide hydrogen, are used to enhance production efficiency, decrease energy levels and improve planning and management. PGRs encourage local fruit production, but also increase the quality and variety of fruit available to marketers and consumers at lower production prices. The production of pear, peach, lemon, strawberry, lithium, apricot, mango and penumin is effectively reduced by SADH/paclobutrazole.

6.7 FLOWERING IN FRUITS

Non-chemical and chemical methods can be used by fruit farmers to cultivate high-quality container-proportional light flowering plants. There are a number of analyses on chemical plant growth regulators (PGRs) which can suppress or promote growth, increase branching, or induce or delay flowering. The Crymidol is Ancymidol may be used for bedding of trees, berries, chrysanthemums, Easter lily, plugs, poinsettia and leafy plants. Most notably, it is healthy for plug plants due to a reduced risk of phytotoxicity. Ancymidol can be used as a foliar spray or substrate drench to monitor plant height, which leads mainly to internal elongation leading to compact plants. In order to continually suspend the chemical in solvents, an ancymidol foliar spray or a drenched substrate solution can be agitated (Davies, 2013).

Lower efficacy at high temperatures may occur, so increased levels of ancymidol may be needed. Phytotoxicity could occur as necrotic (brown, dead) spots at higher application rates. High levels of ancymidol or several low concentrations of ancymidol can also be required in rough plants. Consider evaluating the use of stronger PGRs such as paclobutrazole, flurprimidol or uniconazole in order to boost results and economic efficiency if exceptionally high concentrations of ancymidol are needed. The floral phase of pineapple is controlled by ethylene. Pineapple flowering has been caused by acetylene, calcium carbide, ethephon and NAA (10-15 ppm). The application of paclobutrzol (cultar) soil @ 5 g per tree is successful in the control of mango fruiting. NAA substitute for girdling in litchi for improved flora. SADH encourages the production of grapes, pears, peaches and blueberries. CCC reacts with improved flora to grapes and lemons. Some other examples of growth regulators used in fruit flowering include:

Benzyladenin (BA)

Benzyladenin (BA) can be added to trees, cacti, schlumbergera sp., coneflower, herbaceous perennials, and hosta sp. among other perennial and tropical containerized seeds. Benzyladenin (BA) can be added to bedding plants. Benzyladenine can be used to improve lateral branching and flora and monitor plant height as a foliar spray or as a substratum drench.

Comprehensive coverage is critical as BA has limited manufacturing flexibility.

Chloride in chloramide

Azaleas, bedding plants, geraniums, hibiscus, poinsettia, osteosperm and other herb and wood herb plants can also be treated with chlormequat chloride. Chlorinequat chloride can be used as a foliar spray or as a substrate drench to track plant height, which leads to compact plants. Applications have the most influence when rapid stem elongation begins. The use of foliar sprays often leads to phytotoxic chlorosis (yellowing) but these results can normally be masked by new growth. This is not a cheap control choice, even though it is registered as a drench.

Daminose and Daminocytes

Daminozide instead of lilies can be used to prepare azaleas, bedding, chrysanthemums, gardenias, hydrangeas, poinsettia, other flora and leafy plants. Daminozide should only be used as a foliage spray, often as an internal elongation that produces compressed leaves, to control the height of the plant. The substrates are bound by organic matter in the substrate. Daminozide is highly useful for controlling the height of the plant plug and is most effective in colder climates such as Michigan.

Flurprimidol Primidol. Applicable for beds and plugs, bulb bulbs, chrysanthemums, geranium, poinsettia, herbaceous gardens and other foliage or flowering plants. Flurprimidol may be used for this purpose. Flurprimidol can be added to track plant height leading to compact plants, such as foliar spray or substrate drenching. Note that flurprimidol is not recommended for use in begonium fiber. Due to the application of foliar spray to large quantities of flurprimidol and high temperatures, Vinca can cause black leaves to be spotted. Gibberellins such as GA3 can be used to make tree types using azalea, chrysanthemums, fuchsia, geraniums, lantanas and poinsettias. Gibberellin is also introduced as a foliar spray. It can be used to promote the development and reverse the overdose of plant growth, which lowers PGRs.

Growth regulators are also used n inhabiting flowering in fruits. This is in the case where the weather conditions experienced are not favorable for the plant to flower properly. Inhibition or pause of flowering is done in fruit such as apples, pears, figs, raisins, etc. The application of the GA constitutes parthenocarpic fruit. Parthenocarpic Collection of Cytokine Grape Berries.

GA raises the volume of fruit such as pineapple, strawberries, raspberries and cherries. This implies that growth regulators applied to facilitate flowering can also be applied in hindering plant flowering (Davies, 2013).

6.8 FRUIT THINNING

The use of PGR for the dilution of fruit is common practice in fruit business tactics. That they should have overcome the big challenges of handling apple producers. This might ensure the viability of the fruit company. It is very difficult to determine the correct dilution procedure since it depends on the form and season of the olive tree. In this way, next year, fruit yields will be decreased and the plants will stop flowering. Cytokinin 6-benzyladenine (6-BA) and auxin NAA and NAD, ethylene-released agent Ethrel/ethephone, and carbaryl, a non-PGR compound currently recorded for post-bloom apple thinning in the US. SES goods are also added in separate versions during the thinning window. Any of these pesticides are subject to very stringent export restrictions. Retailers such as Whole Foods and Eco-Apple (for local farmers selling Whole Foods brands) would not take carbaryl-treated apples (Valpuesta, 2002). The scope of thinner people is shrinking, and research into new options is becoming a focus. In our tests, two choices were presented: including current products and the study of new chemicals, such as ACC and metamitron. 6-A-NAA, B.A. The mixture of 6-BA+NAA was as solid as the usual carbaryl mixture used in agriculture. This is a schematic of the many thinner items tested on the old Gala/M.9 block with a fruit scale of 10-12 mm or 18 mm. Both drug combinations provide a substantial reduction in seed load relative to untreated plants. The 6-fold composition of fruit juice was slightly stronger than the dilution of fruit with 100 ppm 6-BA and 10 ppm NAA (PF and 10-12 mm). The narrow fruit harvest from the 6-BA+NAA combination meant that it still offered decent efficiency relative to the NAA combination. There is the application for apple dilution after bloom. DNOC in stone fruit (sodium 4,6-dinitro-o-cresol). GA Preflower for the perfect set of berry clusters, loose and enticing grapes. Post-bloom CPPU therapy is a cytokinin derivative that increases berry size in Kiwi.

Fruit development and ripening: avoidance of fruit drops: NAA, 2,4-D and 2,4,5-T for mango and citrus drops. Fruit maturation: Etherel used in apples for uniform maturation and precocious fruit maturation. Citrus etephone guarantees preservative grading during harvesting. Submersible yellow pigment in 1000 ppm ethephon in lemons. For the maturation of pineapple, mango, etc (Valpuesta, 2002).

6.9 HARVESTING OF FRUITS

Ethylene is considered a 'ripening' hormone. The use of exogenous ethylene and ethylene growth inhibitors (e.g. amino ethoxyvinylglycin or vision (e.g. 1-methyncyclopropene) to regulate post-hare product behavior) to control fruit and vegetable manipulation is common. However, additional plant growth regulators can also be used in horticultural production and have implications for post-harvest efficiency. Many plant growth regulators are on the market in the production of stone fruit. The auxins are used to maximize the size of the fruit. The scale of the fruit and the firmness of the cherries are increased by gibberellin. In addition, when peaches are processed at the end of stem, gibberellin can slow down storage problems, including internal browning and growth of wool. Synthetic gibberellin inhibitors that avoid shooter elongation have also been used to increase fruit size by reducing fruit competition to vegetation.

It has also been shown that these plant growth regulators reduce fungal infections by slowing or senescence or by modifying the composition of the fruit. Other esteemed plant growth regulators have been employed to regulate the biotic and abiotic tension of fruit and vegetable harvesting. Salicylic acid and jasmonic acid activate antioxidant reactions to suppress mushrooms and cause susceptibility to damage at low fruit and vegetable temperatures. Growers and researchers should be aware of the benefits of post-harvest stabilization that field-built plant growth regulators can provide. Spray for etheric goods produced by harvesting walnuts, pecans, olives, cherries and clusters. Increased fruit consistency: GA3 for loose fruit clusters, lower fruit yield, decrease the sum of berries per cluster and increase the size of the remaining berries and the size of the grape. GA3 (50-100 ppm) (25-50 ppm). Dip the grapes in GA3 (75 ppm) for 10 seconds to change the scale of the grapes (Shri Mohan Jain & P.M. Priyadarshan, 2008). There are various examples of how growth regulations are used in different fruits.

Apples

Order of Apple Plant Production

Plant growth regulators, Apogee or Kudos 27.5 WDG inhibitors of gibberellin synthesis (Prohexidione-calcium). Trees of Apogee or Kudos 27.5 have the same number of shoots as untreated trees, but they are thicker or wider in diameter and shorter than those of a compact pupil. The WDG of Apogee or Kudos 27.5 does not decrease the number of leaves or the size of the fruit.

WDG may: when correctly used in apple orchards, Apogee or Kudos 27.5; Reduce shooting development by 20-60 percent and reduce sleeping plumes and/or summer plumes.

Contribute to red cultivars to improve the color of the berries. The result is a tree canopy that maximizes spray coverage. The impact and severity of fire burning on shoots is minimized by minimizing growth and rendering trees less vulnerable to disease shooting. Application: terminal growth, fruit set and Apogee/Kudos 27.5 WDG reactions differ from one field to another. The submission rate and the calendar date can also vary from region to region. If the terminal shoot(s) are shorter than 2.5-7.5 cm or less, first extend the late flowering to the early petal decline. Precise timing is critical, and early use (2.5-5 cm) is more effective than later stages. For Apogee or Kudos 27.5 WDG to be incorporated into the leaf, a sufficient leaf area shall be available. WDGs are non-toxic to Apogee or Kudos bees 27.5. The rate of application is largely determined by height, severity and whether the purpose is to defend the tree from the scourge. Fire blight diseases should be prevented from this stage forward. Follow the symbol to adjust the tree-row dilution values. Projected apogee or Kudos 27.5 of the WDG Tariffs 275 gives information on spray rates of 1 000 L per hectare (dilute) (g product per ha). Repeat requests should be made between 14 and 21 days based on the amount of growth control required. The Apogee or Kudos 27,5 WDG lasts about 14 days and falls into the trees for a few weeks in order to slow down its growth. There could be two ways to conduct accurate surveillance. More vigilance could be required in areas with a long growing season, highly tolerant trees or fruitless trees. No more than four applicants are valid for any season. Do not add more than 5.4 kg per hectare of land per season to Apogee or Kudos. Not valid within 45 days of harvest. Apogee or Kudos 27.5 WDG is readily used in more extreme sprays as long as the canopy is fully wetted. It is not recommended to spray low-volume plant growth regulators and chemical thinners.

The values of WDGs 27.5 shall be balanced by the orchard and the environment in terms of the Apogee or the Kudos. See Table 2 Apogee or Kudos 27.5 WDG Orchard and Environmental Requirements and Spray Number for more information.

Repeat applications should be made between 14 and 21 days based on the amount of growth control required. Apogee needs about 14 days to slow down production. After a few weeks, it breaks down in the ground so two applications for seasonal growth control can be expected.

Encourage the return of trees for the coming season

- Never again (L-fruiting): NAA is a plant growth regulator used for the development of an auxin plant that can effectively facilitate return flowering when used at the thinning point in two-year-old plants. For biennial trees a year with full flowers and hard fruits, it can be called "on-year" and "off-year" when flora and fruit are lightly planted. The NAA application will be the most successful in the "on-year" scenario for the next year, promoting the return of flora during the "off-year"
  Apply water from full blossom to 30 days from full blossom at a rate of 39-310 mL/1000 L (1.2-9.7 ppm). Use significant quantities of water in slow drying conditions. After 2-3 hours, NAA is considered dull. You need to pair NAA with carbaryl (Sevin XLR). Hot, sluggish and slow drying are the best conditions for pumping NAA, often in the morning. It is safest to send between 21°C and 24°C. This chemical should not be used at or above 15.6°C or above 26.7°C. This chemical is to be used.
  PARATION: should not refer to pygmies in cultivars prone to cultivation in the same growth cycle as MaxCel or Cilis Plus (example. Fuji, Red Delicious, Golden Delicious and Gala). Improving the flowering of inexpensive trees
- Etherel's: Ethrel (ethephon) can be used in non-bearing plants to support flora.
  Apply Ethrel foliar spray to unsticky apple trees 1 to 2 weeks after flowering (determined by bearing trees in the area). Delay applications to prevent over-shining and malformed fruit in young orchard trees from beginning with a few flowers until 3 to 5 weeks after flowering. Trees should be big enough to grow fruit at a time of year. Mix 2 liters of Ethrel in 1,000 liters of water for spur-type trees and use it for diluted spray at runoff level. 4.25 liters of Ethrel in 1,000 liters of water for non-spur plants. For concentrated use, apply 7 L Ethrel per hectare to the spur species or 14 L Ethrel to the non-spur type in 500 L water. Notice that this rate would completely remove any fruit from the trees, particularly if used earlier than 4 weeks after full bloom.
  Encouraging coloring of fruit
- Etherel: Ethrel (ethephon) can be used to render the fruit colors of apples that are sold ripened and grown immediately. Apples

treated with Ethrel have not been processed for a long time, and the fruit would drop early.

Using stop drops like naphthalene acid to prevent unwanted premature drops after use of Ethrel (NAA). Two applications are required: one when Ethrel is used as a tank mixer and five days later. Apply the stop decline to the label level. Spray just the number of trees that can be harvested over a period of 2-3 days. Fruit, depending on the variety and air temperature, will be ready for harvest 5-10 days after spraying. Readiness, starch, flavor, and color until sprinkling every day to find the ripeness of apples. A few trees sprayed for 3-4 days are easier to deal with than a big amount sprayed at the same time. Make sure there is a demand, particularly for early growers, for processed apples before spraying. The rate of application depends on the cultivar, the date of application, the intensity of the tree, the temperature, the weather and the required response rate. The earliest growers such as Jerseymac and Paulared, growing cool, use 0.75-1.5 liters per hectare of Ethrel with ample water to thoroughly moisten the plants. It takes 1.5-4.25 l/h for McIntosh. Increased thresholds for nitrogen-rich trees or poorly pruned trees in the early season. Lower values are used in young, well-preserved trees or low nitrogen trees. The best results are reached with the addition of Ethrel 2-3 weeks before harvesting. Good spray coverage is critical.

- Blush: Blush is used to improve the production and growth of anthocyanin in two-colored apples, Fuji, Gala, Honey crisp and McIntosh. It can also contribute to the solution of delays in the coloring of Hold apples. Blush increases the breakdown of chlorophyll, which leads to reddish berries, anthocyanin secretion and carotene. Color improvements cannot be seen until the strain of the variety is shown to be high in color or until the orchard grows well under climatic conditions.

  Take 1-2 Blush applications at a rate of 190-381 mL/100 L of water 7-28 days before the expected harvest (100-200 ppm). Apply the first treatment 21-28 days before harvest and the second request 7-14 days after both treatments. The first treatment should be made before the red color begins to form. The results vary and can be influenced by environmental and diversity factors. Do not apply to trees that have been affected by excessive sunshine, drought, illness or stress. Using to achieve the best results and avoid doing it at the hottest time of the day under slow drying criteria.

Apple Pre-Harvesting Chemical Control

Apples are subject to variation in cultivars prior to selection. McIntosh and Honey crisp are the most interesting cultivars in Ontario. Signs related to the early drop in fruit include:

High nitrogen content late summer trees with high harvest weights and warm temperatures weeks just before and after harvesting, arbors with insufficient rainfall, boron or magnesium nutrients, trees with severe pest damage, fruit trees are grown in NAA-classified clusters producing large fruit products such as Fruiting-L and Fruit Repair.

Fruiting-L and Fruit Reparation.

When the first undamaged apples start to fall, add NAA at a rate of 10 ppm. This is very helpful timing. Do not mix the usual pre-harvest drop with sound apples that collapse due to insects, disease or nutrient deficiency. NAA is effective for 7-10 days after sprinkling. Repeat application 10 ppm 5-6 days after initial stop-drop application. It will take NAA 1-2 days in warm weather and longer in cold weather. It can also take longer for NAA to take place before the orchard has been picked. Use NAA when the spot set is complete. NAA is internationally systematic only and needs to be applied to allow more precise coverage of dilute mist. High amounts of water take longer to dry.

In conclusion, growth regulators can be used in different stages of farming to achieve results desired by farmers. PGRs may be used with a certain accuracy for thinning fruits such as berries, pears and apples among others, regulating growth and changing harvesting times. These products are absorbed by plant cells, mainly leaves and fruit, which interfere with the plant's biological machinery, by imitating natural plant hormones or by preventing the production of natural plant hormones. Allow farmers to ensure a continuous supply of fruits.

Chapter 7

Fruit Training and Pruning

CONTENTS

7.1 INTRODUCTION TO PRUNING AND TRAINING

Pruning and preparing trees helps produce great yearly yields of value. This training creates a solid system that increases the tree's lifespan and reduces loss from branch breakage. Improvement of product quality depends, to a limited extent, on appropriate pruning. The type and measure of pruning are to a great extent decided by the breed, cultivar, and period of growing trees. Other factors are the structure, tree, and development and fruiting propensities.

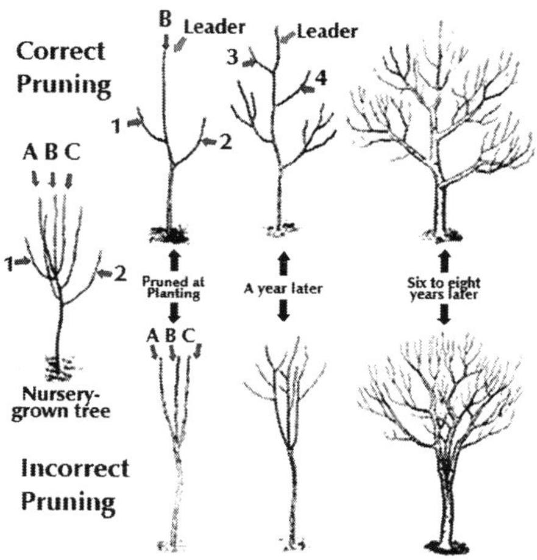

Figure 7.1. Fruit training and pruning in apples

Source:https://www.google.com/imgres?imgurl=https%3A%2F%2Fhgic.clemson.edu%2Fwp-content%2Fuploads-training-apple-pear-trees

Intensely pruning young trees should be avoided since it might defer the hour of bearing from 1 to 3 years. Then again, failure to prune during the initial 3 or 4 years can bring about a poor tree with feeble, thinbranches. Keen preparation methods during the initial few years are basic to build up a strong structure and tree shape. When the basic tree structure is developed, light yearly pruning is recommended until the tree starts bearing. Inadequate pruning of bearing trees may result in little, inadequately grown leaves with a low sugar content also. Trees that are over pruned may additionally produce non-nutritious fruits in spite of their huge size. Such trees typically have huge, delicate cells and thus a short fruiting life. Notwithstanding influencing tree quality, pruning affects the fruiting capability of the tree for the next year (Saleh, 2011).

Tree health, nitrogen treatment, and the type of crop determine the measure of pruning required for the fruiting tree. Extreme pruning and in addition to extreme nitrogen manure bring about overwhelming shoot development with almost no blossom bud development for the following year's harvest. Largescale fruiting, especially related minimal pruning or potentially high nitrogen, brings about a deformed tree that can't shape lower buds. Equilibrium should be kept up between vegetative development and fruiting through the processes of pruning, nitrogen supplementation, and yield. The best technique for deciding the amount to prune is to pass judgment on the tree health and foreseen crop. For most bearing apple trees, the tree is considered to have sufficient health in the event that it creates a normal of 10 to 15 branches of terminal shoot development. In the event that the terminal development is not as much as this, it should be pruned more vigorously than the earlier year or more nitrogen should be applied. On the off chance that normal terminal development is higher than 15 inches, less pruning should be done or then again nitrogen reduced. In case a substantial harvest is envisioned, the measure of pruning can be expanded fairly, however, pruning ought not to be utilized as a substitute for organic product diminishing (Saleh, 2011).

Fruit training is training in which tree development is coordinated into an ideal shape and structure. Training young organic product trees is basic for appropriate tree improvement. It is smarter to coordinate tree development with preparation than to address it with pruning. Training incorporates summer preparing and summer pruning just as lethargic pruning. The objective of tree preparing is to coordinate tree development. Pruning is the appropriate and wise removal of plant parts, for example, shoots, spikes, leaves, roots or other terminal parts to maintain tree structure and increment its productivity.

It is done to make the plant more profitable and bear quality fruits, the increased life span of the tree, make it possible of for the farmer to get the most from their plantation. Pruning is a continuous cycle and can be utilized to keep up any ideal tree size. Removal of a branch eliminates not just excessive starches but decreases the leaf surface size also. Pruning develops fruit product size, nitrogen quotient and development rate.

However, it is important to note that excessive pruning lessens productivity particularly with young trees that may effectively be creating an excessive amount of vegetative development. Large cuts bring about inordinate stimulation of fledglings close to the cut, while all-around trees spread nutrients better over the whole tree. The seriousness, type and measure of pruning to be done on a tree rely upon the age, existing system, state of bark and wood, Growth qualities, fruiting propensity for the variety, whether the tree is long-lasting or a temporary breed. Pruning is frequently done throughout the colder time of year, usually alluded to as lethargic pruning.

Training targets can range from a variety of things such as; to concede more daylight and air to the center of the tree and uncovering the most extreme leaf surface to the daylight. To coordinate the development of the tree so that different social activities, for example, showering furthermore, reaping is performed at the least cost. To shield the tree from sunburn and wind damage and to provide a reasonable circulation of fruit bearing parts on the fundamental appendages of the plant (Saleh, 2011).

7.2 GUIDELINE OF TRAINING

The guiding philosophy in developing a young-looking tree is to create a solid root and branch system. All techniques for preparing should stand or fall by their capacity to develop a tree to bear high yielding fruits without failure. The trunk is the main stem of the plant. The head is the point in the bud from which the first branch emerges. The Scaffold branches which are the Main branches emerge from the head are also known as structural branches. Low headed trees are those which structural branches emerge inside 0.7-0.9 m range from ground level. Low headed trees come into bearing much earlier, can resist turbulent winds all the more adequately also, and their watering and reaping costs are less.

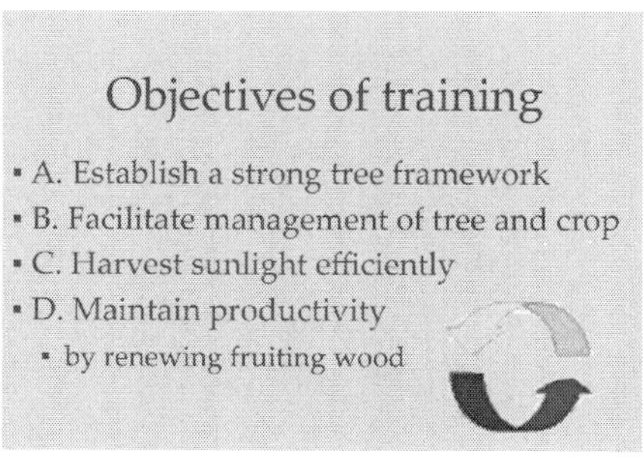

Figure 7.2. Objectives of training

Source: https://www.google.com/url?sa=i&url=https%3A%2F%2Fslideplayer.com%2Fslide%source=images&c

- High headed tree: Trees in which platform branches come out from the storage compartment above 1.2 m. In the tropical atmosphere, high headed trees are inadmissible as their uncovered trunks are exposed to sunscald in summer.
- Crotch: The point made by platform appendage to the storage compartment or the auxiliary branch to structural appendage is called crotch. The crotch should be wide and not thin.
- Leader: The primary developing branch from ground level up to the tip overwhelming any remaining branches.
- Spur: Numerous shoot development which is bountiful over the organic product trees and whereupon a large portion of the organic product is borne. Water shoots are remarkable fiery vegetative shoots that develop from the high focuses on the fundamental branches upstanding way to the detriment of principle branches (Varkulevicius, 2010).
- Suckers: emerge from unique buds on the roots or underground pieces of the stem of the tree.

7.3 IMPORTANCE OF PRUNING

- To control the size of the plant.
- To control the structure (underlying makeup of the plant) which includes number, situation, relative size and point of branches.
- Better quality natural products by better light dispersion.
- To eliminate ailing, befuddled, dried and broken branches.
- To eliminate the non-beneficial parts to redirect the energy into those parts that are equipped for bearing organic products.
- Proper extent of root-shoot proportion.
- To direct the organic product crop.
- Longevity of the tree.
- Chances of bugs nuisances, infections and winter injury are less.

Pruning is done in order to, allow more daylight, eliminate ineffective branches which are delivering less or no foods are grown from the ground, to keep the plant in its appropriate state, essentialness and to acquire ideal yields of great quality natural products.

7.4 TYPES OF PRUNING

- Dispersing: When a shoot is altogether taken out from the plant and no re-development is permitted to happen from the cut closures don't stimulate the tree.
- Heading back: When the terminal segment of branch/shoot is taken out and it energizes sidelong development from the excess shoot. All in all, if a bit of a shoot is eliminated also, the development can proceed from the excess parts. Heading back advances the development of lower buds just as a few terminal buds beneath the cut. When sidelong branches are going into one-year-old wood, the region close to the cut is fortified. The headed branch is a lot more grounded and inflexible, bringing about parallel auxiliary spreading.
- Meager wood pruning: alludes to the evacuation of moderate developing, powerless, under hanging branches or shoots which are either not fruiting or delivering products of inferior quality.
- Bench Cut - eliminates lively, upstanding shoots back to side branches that are moderately level and outward developing. It's used to open up the focal point of the tree and spread the branches outward. This is a significant cut and should just be utilized at the point where necessary (Varkulevicius, 2010).

7.5 SEASONS OF PRUNING

Figure 7.3. Best Time to Prune

Source:https://www.google.com/url?sa=i&url=https%3A%2F%2Fwww.gardeningknowhow.com%2Fedible%2Ffruits%2Ffegen%2Ffruit-tree-pruning

Most orchardists prune during the slow season. During this time, troublesome branches can be all the more effortlessly identified. Slow growing trees are not without any problem debarked by jumping on the branches and the bark only occasionally pulled from pruning wounds. The ideal time to prune is in late-winter, not long before the start of dynamic development. Around then, injuries recuperate most promptly and blossom buds can be without any problem identified. This is particularly valid for peach trees. Notwithstanding, the pruning of huge plantations must be begun before so employment can be done before the trees continue development. Low temperature damage is a significant risk with fall or late-fall pruning in Iowa. Trees that have been pruned before extreme climate are frequently genuinely harmed by the freezing temperatures that follow (Riccardo Gucci & Claudio Cantini, 2000).

Pruning in November and December can be more dangerous to trees than pruning in late February, March, or early April. Summer pruning isn't for the most part suggested since it causes more overshadowing of the tree than slow season pruning. On the off chance that a predominating impact is wanted, at that point some midyear pruning might be needed. Water fledglings might be taken out in June and July, when 6 to 10 inches in length, either by hand or with shears. Evacuation of water sprouts as of now keeps the tree open for good light entrance and lessens the measure of slow pruning required.

Pruning Equipment

For the small-scale orchardist or homebased cultivator, hand instruments give the best and most efficient pruning hardware. Hand shears, trimming shears, post pruners, and pruning saws of different sizes and styles are accessible. Each apparatus will perform well assuming appropriately utilized and the forefronts are kept sharp. Two-legged or on the other hand three-legged plantation stepping stools help in coming to the highest points of taller trees. Corporate orchardists have adopted different sorts of intensity worked hardware to make the pruning activity simpler, faster, and more proficient. Most power pruners work on the rule of packed air or water powered liquid actuating a cylinder that works the shear head. Water driven controlled chain or round cutting edge saws are likewise accessible. Small cutting tools are likewise utilized in plantation pruning. These saws are particularly helpful when enormous cuts should be made (Riccardo Gucci & Claudio Cantini, 2000).

Alongside the utilization of intensity, pruners have come with the utilization of different sorts of stages and portable towers that move laborers through the plantation. Such gear is economically accessible, albeit a few producers have constructed their own. Advancement and use of such gear has expanded the effectiveness of power pruners, fastened pruning in enormous plantations, and made the whole occupation simpler by getting rid of awkward stepping stools or the need to physically climb trees. The utilization of pruning stages and water-driven pinnacles makes it possible to prune a tree from an external perspective around the middle. This is a productive and simple way to prune the external and higher branches. All the light may enter the inside of a tree, instigating more gainful wood growth around there. The net outcome is a more noteworthy presentation to the light of the whole leaf and fruiting territories of the tree, which builds the creation of well colored, great natural product. Use power pruners with care since over-pruning and ill-advised cuts can happen without any problem.

Supporting machines can be utilized in deciduous plantations. A supporting machine does pruning in a short timeframe. Supporting isn't a strategy for specific pruning; subsequently, plantations pruned this way actually need some nitty-gritty pruning by hand or on the other hand power shears. Whenever nitty-gritty pruning doesn't follow supporting, the trees will become thick and shaggy in the external regions. At the point when this occurs, a weak tone will be delivered in the inside of the trees. Supporting together with a decent yearly pruning program can bring about significant time and work investment funds, particularly in enormous plantations.

Making the Cuts: Two essential cuts are utilized for pruning organic product trees-the "dispersing" cut and the "heading-back" cut. The cut utilized relies on the age and area of the wood and whether the purpose is to prune a course tree or train a young tree. The dispersing part is used to remove a whole shoot or branch. It eliminates undesirable, 1-year-old shoots while choosing an essential structural arrangement also, optional branches. It is utilized to kill the "crow's foot" disease that, assuming left, can bring about the stifling out of the essential leaves and the destruction of plant evenness. Other undesirable shoots are those developing on the top or base sides of a platform. Shoots on the top side of a platform, for example, water sprouts, are enthusiastic and will in general stay energetic with little potential for fruiting. They conceal out lower partitions of the tree, which brings about a decrease in organic product tone also, potential organic product bearing territory. In the event that these branches do produce organic products, they become enormous, subject to appendage rub, and have helpless capacity quality. Shoots emerging from the base side of a platform are powerless from concealing and produce little, ineffectively covered organic products (Riccardo Gucci & Claudio Cantini, 2000).

Wanted shoots are those of moderate life that emerge from the sides of the horizontal platforms. On developing, bearing trees, some dispersing of branches can be drilled to lessen stuffing, yet care should be taken not to leave exposed territories along with the platforms.

The heading-back slice is utilized to eliminate a bit of a shoot or branch. At the point when this cut is made on 1-year-old shoots, it supports fiery shoot improvement close to the cut. In preparing youthful trees, this cut is utilized on 1-year-old shoots to support sidelong stretching on the focal pioneer for future platform advancement. On platform pioneers, the heading back of 1-year-old shoots is utilized to keep the framework from turning out to be frail and dainty. This is particularly valid for cultivars that have a spreading or dainty development propensity. Abuse of the heading-back cut on 1-year-old wood will postpone fruiting. For bearing trees, a comparative cut is utilized on more seasoned wood to contain development, lessen swarming, or revive old prod wood. Quite a cut is alluded to as "going to a parallel." This cut into more seasoned wood doesn't deliver the strengthening reaction related to the heading back of 1-year-old shoots.

Summer heading back or squeezing of momentum season's development is an elective practice used to control force. This cut creates a predominant impact by animating powerless regrowth on the headed shoot. Lively undesirable shoots can be squeezed back whenever from June through August. Notwithstanding, on apple and pear cultivars that are vulnerable to fire curse, this practice should be postponed until July when fire scourge contaminations are not, at this point dynamic.

All cuts should be made with sharp, effectively changed apparatuses that give perfect, smooth cuts. In dispersing a shoot or eliminating a branch, the slice is made near also, corresponding with the supporting appendage. When heading back to a sidelong, place the cut so it proceeds in the general line of course of the horizontal branch. Stubs try not to recuperate and may begin rot. In the event that the cut is as well near the purpose of the horizontal connection, the branch is liable to part out. Likewise, on platforms, going to a horizontal that starts from the top side of the appendage beats the propensity to part out.

At the point when shears or loppers are utilized, the cutting edge is set under the branch to be taken out and against the supporting appendages to take into account a smooth, close cut. The wood on the underside of a branch is gentler so there is less danger of springing the shears. The cut can be made all the more effectively by pushing marginally against the branch toward the path the cut is being made. Never squirm the shears through a cut as this makes a worn-out injury that will mend gradually, and may spring the lances. In eliminating an enormous appendage with a saw, it is frequently useful to do as such with three cuts. Initial, an undercut 1 to 2 inches deep is made around 8 to 10 crawls from the base of the appendage. At that point a cut from above is made, beginning 1 to 2 creeps past the undercut. Along these lines, the appendage will fall without destroying the bark into the supporting appendage. At long last, the short stub is eliminated with a nearby cut made against the supporting appendage (Riccardo Gucci & Claudio Cantini, 2000).

7.6 DORMANT PRUNING

Frequently done throughout the colder time of year ordinarily alluded to as lethargic pruning. Dormant pruning is considered to be an animating cycle. Heavy slow pruning additionally advances unreasonable vegetative life, and the timing of slow pruning is basic. Pruning should start as late in the colder time of year as conceivable to stay away from winter damage.

Reasons why Dormant Pruning is considered:

- Visible Branching Patterns (branches can be seen effectively right now).
- Other plantation activities are less squeezing
- Less threat of pulling bark away from around pruning wounds
- Pruning works out in a good way for a cool climate.
- Increased development reaction
- Has less predominant impact on the trees than summer pruning.
- Improved Equipment Access
- Fungus illnesses are lethargic as well, so there's less threat of transmission
- Tree is going to begin the dynamic development that advances wound-recuperating

7.7 SUMMER PRUNING

Done during summers-alluded to as summer pruning. It's less serious, more uncommon, more explicit and particular. It eliminates an energy or food-delivering bit of the tree and results in decreased tree development. For most purposes, summer pruning should be restricted to eliminating the upstanding and energetic ebb and flow season's development; just diminishing cuts should be utilized. To limit the potential for winter damage, summer pruning ought not to be done after the finish of July.

Why most fruit farmers prefer Summer Pruning:

- Better brightening of leaves and fruits.
- Improved osmosis in the focus of trees
- Better natural product tone, organic product quality and capacity.
- Slower complete development of trees
- Less wood creation
- Less complete pruning exertion and improved work proficiency
- Possibility of decreasing crown volume of the tree, better use of room, more tree per section of land
- Less risk of injuries that occur during winter.

7.8 TRAINING SYSTEMS OF FRUIT TREES IN TRAINING AND PRUNING

- Focal Leader System: The main trunk stretches out from the soil surface to the aggregate tallness of the tree. Several side branches develop at various statures in different bearings. The advantages of this system is: Such trees are basically the most appropriate to bear crop load and to oppose the harm from solid breezes. On the other hand, the disadvantages include: Trees under this framework develop too tall and are less spreading. Tree the executives (splashing, pruning, diminishing and gathering) is troublesome. Shading impact on inside covering (the lower parts of such trees might be such a huge amount in a shade that the natural product may not be capable to create a legitimate tone).
- Open Center System: This is where the main trunk is permitted to develop up to 1.0 m by cutting inside a time of planting. 3-5 sidelong branches are permitted to create from a short principle stem. Widely utilized for peaches and is useful for mechanical

collecting. The advantages of this system include: The trees so prepared permit the greatest daylight to come to their branches. The better coloration of organic products on the inside side of the tree. Trees are more productive and low-spreading tree incredibly encourage tasks like showering, pruning, diminishing and reaping. The disadvantages are: Such trees are fundamentally feeble, and their appendages are more prone to break with crop load and solid breezes. This framework doesn't just need serious pruning begin with yet additionally steady exertion to keep up its structure through extreme pruning treatment (Saleh, 2011).

- Changed Leader System: This preparation framework is generally satisfactory for corporate pruning. This framework consolidates the best characteristics of the focal pioneer and open focus frameworks. A pioneer creates on the youthful tree until it comes to the stature of 2-3 m and afterward the development is confined. Laterals are chosen to branch in a twisting design up the focal trunk and are cut until the legitimate number and dissemination of branches have been acquired. The advantages of these systems are: The branches are very much dispersed, permitting a lot of daylight to arrive at the inside of the tree. The trees are primarily solid and not inclined to appendage breakage. Owing to the restricted stature of trees, splashing, pruning and collecting might be done without any problem. Some organic product trees bear on flow season's shoots while others do as such on the past season's development. The bearing may likewise be on prods and additionally shoots. Mango, grape and most calm natural products have bloom buds shaped significantly sooner to the appearance of their blossoms, however, citrus organic products have bloom buds separated only a little prior to blooming.
- Some trees have straightforward blossom buds while others have blended buds—which after unfurling give blossoms just as leaves. The pruning practice should be synchronized with the blossoming propensity for the species and once in a while indeed, even the assortment. Evidently, one should know about the qualities of the plant material close by in any case there are odds of delivering the trees less beneficial and here and there out right desolate through aimless pruning (Saleh, 2011).

7.9 FRAMEWORKS OF TRAINING DWARF TREES

- Shaft Bush: adjustment of the bantam pyramid or as the middle of the road between a vertical cordon and a hedge structure. Differs from the bantam pyramid in that it has no particular game plan of framework branches and from the vertical cordon in that the tree is borne on short branches as opposed to straightforwardly on the principle stem or trunk. The most significant component of this framework is the securing of sidelong shoots in a flat situation with almost no mid-year pruning. Trained with or without help posts with a focal pioneer straight and with numerous little fruiting branches. These branches are bowed out and somewhere around spreaders to grow wide groins and to initiate early fruiting. Tree spread is constrained by scaling back the shoots to ½ to ¾ of their length or back to frail laterals.
- Bantam Pyramids: comprises of a tree with a focal stem about 2.5 m tall from which short branches transmit in progressive levels so a pyramidal shape is fabricating up. Fruiting prods are created on the short branches. Summer pruning structures a basic part of progress with bantam pyramids.
- Cordons: are single-stemmed trees. This structure is rarely found in business organic product creation and has been supplanted by bantam pyramids. Cordons are vertical, slanted or level Vertical cordons: trees are prepared erect and developed to a tallness of 10-12 feet with the first fruiting wood created at around 30 cm. starting from the earliest stage. Slanted cordons: are prepared at the 450 points (Varkulevicius, 2010).
- Palmettes: alludes to the state of a palm leaf, an open fan or an open hand with spread fingers. Formed by creating sidelong framework branches from the earliest starting point at 30 cm from the outside of the ground. These laterals are two by two and are similarly adjusted in inverse ways. Other frameworks are thus vivaciously created, be that as it may, are kept around 30 cm separated. Four gatherings: Horizontal, Oblique, Candelabra, Fan Palmette.
- Espaliers: like the knifing arrangement of preparing grapes. The lattice is 5 feet high with one or the other 2 wires at 3 feet and 5 feet OR 4 wires at 2,3,4, and 5 feet.
- Hedgerow Plantation: Adapted to mechanical development.

Distance inside the lines - close and between the lines - wide to oblige apparatus and the board. Adapted to minimal trees, marked unattached on trellised developing north to south. Trees planted genuinely close 4 feet in lines, 12 feet separated, what's more, developed as little thick trees that form into a constant or hedgerow propensity. Hedgerow ranch can likewise be created without uphold utilizing semi midget or semi-standard rootstocks with trees 10-15 feet separated in lines 20-30 feet separated. Trees are kept 10 feet in tallness.

- Glade or Bed Orchards: In this framework, trees are planted at ultra-high densities of 30,000 to 100,000 trees for each hectare. Trees dispersed 30 x 45 cm separated with sprinkler head through which hormones, manures, herbicides, pesticides and water might be applied. Cropping is taken each other year up to 20 t/section of land yet trimming each year might be conceivable by wise pruning and the board. It offers total motorization of both pruning and collecting combined with substance control of weeds. However, there are a few restrictions and they include: High expense of foundation. Orchard becomes uneconomical previously standard plantations. Annual creation is relinquished on the grounds that of the expulsion of fruiting buds in substitute years (Varkulevicius, 2010).
- Tatura Trellis framework in Peaches: One of the most elevated yielding frameworks. The system has all the earmarks of being V formed. Trees are planted around 6 x 1 m separated (1668 trees/ha). Bearing beginnings in the subsequent year. The main structure comprises Y-formed pioneers that are attached with the wires.

7.10 PRUNING BEARING FRUIT TREES

Pruning to keep up tree shape and size is fundamental for the duration of the life of a tree. Notwithstanding upkeep pruning, youthful bearing trees do require some branch pruning to increase light infiltration. The initial phase in pruning an orientation tree is to eliminate any enormous, undesirable appendages and dead, unhealthy, or harmed branches. This should be done every year, despite tree age. In like manner, minimize water provision every year aside from an infrequent one that is required for growing new bearing surfaces. The most ideal opportunity to eliminate water sprouts is in late-spring at the point when they are delicate. By now they can be effectively focused on with a gloved hand. Whenever done after fledglings become woody at the base, shears will be important to prune them off. Whenever eliminated in the late spring, water sprouts only sometimes develop back, or on the off chance that development happens, it will be little. Any water sprouts staying on the tree can be eliminated during slow pruning. Suckers emerging from the rootstock ought to likewise be taken out yearly (Riccardo Gucci & Claudio Cantini, 2000).

As trees become more established, an expanded number of heading back cuts into more established wood are expected to look after tree size and shape. Additionally, the measure of branch cutting may be expanded to accommodate the spread of wood all through the tree.

Shading of the product created the past season is additionally a significant manual for the measure of pruning required. Poor shading will, by and large, show a requirement for more definite pruning, especially of the dispersing type. Weighty pruning, nonetheless, may bring about huge, inadequately shaded organic product. This is likely if nitrogen applications have been predominant. The pace of nitrogen application should be diminished whenever pruning is heavier than typical. Yearly pruning is suggested. When followed, just a moderate sum is important every year. One objective is to eliminate branches that are harshly concealed and will bear not many fruiting prods. Eliminate branches that twist to the ground or head back to upward developing laterals. Yearly lethargic pruning of bearing trees serves to advance normal bearing. It likewise will in general decrease organic product over-burdens, just as the work consumption for hand diminishing. When pruning is unpredictable, trees become as well thick and frequently biennial in bearing.

Apples

Apple cultivars vary in development and fruiting propensity. A significant bit of the blossom buds are borne terminally on prods with a minor bit being delivered terminally on shoots. Now and then cultivars will deliver blossoms horizontally on 1-year-old shoots. This implies it is prudent to think about the development propensity for every cultivar when creating pruning strategies. N1 shading strains or transformations of the guaranteed cultivar are pruned equivalent to the parent cultivar. Over color of the products of the all-red changes is great, yet legitimate pruning is basic to improve natural product size.

Cortland

Cortland trees will in general create a significant part of their blossom buds terminally on short shoots, numerous of which should be eliminated yearly. This requires itemized pruning. Expulsion of slim, underhanging branches is significant as these produce inadequately hued foods grown from the ground conceal different branches and natural products. Bigger branches might be dispersed by making a few little, uniformly conveyed cuts, starting at the finish of the branch and working to its base.

Delectable

Delectable and its different strains have a characteristic inclination to create slender groins. This is especially valid for the prod type strains. Such groins as often as possible create bark considerations that add to a debilitated tree structure. Along these lines, spreading, reshooting, and restorative slices to space the best branches along the storage compartment of the trees are significant systems during the second to fourth years. Since the prod type strains have an extreme propensity to create tight groin points, the utilization of the heading-back cut on 1-year-old platform pioneers is of faulty legitimacy.

One option is to head the key plant parts until the essential system is grown, at that point end utilizing the heading-back cut on 1-year-old wood. The subsequent option is to keep on utilizing the cut on the horizontal chiefs and in the mid-year eliminate the new shoots that create just underneath the new pioneer when they are around 4 to 6 inches long. Meanwhile, deciduous trees may grow an excessive number of medium-sized branches, coming about in over dense trees, except if a calculated evacuation of overabundance branches is conveyed out. Since this cultivar is planted essentially for eating, adequate yearly pruning is important to deliver a high level of alluringly hued apples. This implies that trees should be kept sensibly open to allow passageway of daylight. Eliminating little width and underhanging branches from both the inside and the fringe of the tree will permit light to enter. Such isn't the issue with spike type.

Scrumptious trees, which may require next to zero yearly pruning once they come into bearing. Heavenly has to some degree unstable fruit setting propensity. Point by point pruning may improve the arrangement of organic product. Natural product diminishing is significant with deciduous trees to secure a high level of enormous, all around shaded organic product, especially if pruning has not been done appropriately. Trees are likewise inclined to create a plenitude of water sprouts. These are best eliminated as they grow right off the bat in the developing season.

Golden Delicious

This cultivar is an early carrier and is inclined to produce bloom buds along the side on 1-year-old shoots furthermore, on recently framed spikes. Thusly, tree structure should be created as ahead of schedule as could be expected under the circumstances. It isn't hard to build up a solid, all around formed framework on this cultivar. At the point when they initially come into creation it is shrewd to thin the branches so that the remaining organic products are limited to territories of the tree where the platform is in any event 3 years of age, or the lower half of the platforms. In the event that this isn't done, the wanted state of the system might be lost. Golden Delicious tends to create frail, slender calculated groins, however significantly less so than Scrumptious. It is, thusly, important to build up a system with wide-calculated groins. Creating branches regularly become long and may break with the first weighty harvest of natural product in light of the fact that the wood of Golden Heavenly is somewhat fragile. Pruning every year and keeping all branches inside a sensible length is basic.

Jonathan

These trees have a spreading delicate development propensity furthermore, normally build up a thick, fine structure. Except if appropriately pruned, bearing trees become thick and concealing happens all through the tree. Due to its propensity to deliver numerous blossoms along the side on the shoots, Jonathan doesn't react well to heading-back cuts. Point by point, slender wood pruning and branch separating are fundamental for acceptable natural product size and shading all through the tree. This cultivar will in general be little fruited. Assuming great, popularize natural products are to be delivered, cautious consideration should be given to pruning rehearses. Weighty pruning of Jonathan expands the risk of blight, an infection to which the cultivar is very defenseless. It might likewise lessen natural product tone. Nitrogen compost should be limited on Jonathan all together to improve organic product tone and to diminish fire blight susceptibility. The heavier the pruning, the lesser nitrogen should be applied.

Lodi

Lodi is comparable in development capacity to Yellow Transparent also, both should be pruned comparatively. This cultivar will in general set organic product intensely in alternate years, particularly on the more modest wood in the focal point of the tree. Organic products delivered in the inside are typically of substandard size, green in shading, and average quality. Yearly, point by point pruning in the inside segment is essential to eliminate the vast majority of the meager wood that produces the sub-par natural product. Branches in the external parts of developing trees additionally need yearly dispersing and a few heading back of long branches.

McIntosh

The McIntosh cultivar normally creates framework branches with sensibly wide-calculated groins, bringing about a spreading-type tree with numerous optional branches. In this way, impressive branch diminishing is important to allow careful splashing. McIntosh will in general create natural product spikes all through the tree, like Golden Delicious. Pruning to keep the tree open is critical to creating a very much shaded organic product of good size.

Rome Beauty

Trees of this cultivar produce a significant part of their blossom buds terminally on shoots in contrast to spikes and are portrayed by numerous thin branches that cause a lot of concealing. These branches require significant dispersing. Every enormous branch is best pruned as a unit from tip to the base. The cuts are appropriated as equitably as could be expected under the circumstances, and feeble wood is eliminated. A typical error when pruning bearing Rome Beauty trees is to begin at the base of a huge branch and prune outward, eliminating all the bearing surfaces for a significant distance. This leaves an extended length of desolate wood with a group of unpruned, thick twigs at the edges. In the event that the full harvest potential is to be picked up, exceptional consideration should be given to the situation of dispersing cuts on all strains of Rome Beauty trees.

Yellow Transparent

These trees produce numerous branches and may turn dense. At the point when the primary branches are very much dispersed, the weight of the natural product spreads the tree sufficiently. Almost no going to outward developing laterals is required if enough enormous branches are taken out in pruning. Little cuts, all around circulated on the branches that are left, help impressively to improve organic product size. The measure of feeble wood pruning essential is expanded as trees become more seasoned. Yellow Transparent additionally is extraordinarily profited by early organic product harvesting. Since it is hard to sell little fruits, broad pruning and diminishing are required to improve the organic product size of this cultivar and help it in conquering biennial bearing. Over-pruning should be maintained and a strategic distance since it stimulates deciduous development that is very vulnerable to the fire curse infection.

Pear Trees

Pear trees have similar fruit bearing propensities as apple trees, with the significant segment of the flower buds created terminally on spikes. The measure of harvest created along the side on shoots is significantly less than that happening on apple trees. Light pruning is rehearsed on bearing pear trees. Indeed, even moderate pruning may initiate the advancement of water fledglings and quickly developing terminal development. This sort of development is truly vulnerable to fire blight infections. In all cases, slices should be bound to appendages of a more modest width. Heading back of terminals should be done distinctly as the tree turns out to be excessively high. Going to a sidelong branch around then ought to be light and cuts made into more modest distance across wood. It ordinarily isn't important to prune pear trees each year on the grounds that new development ought not be hefty or on the other hand incredible. Notwithstanding, every year trees should be inspected to decide whether pruning is important.

Eliminate Fire-Blighted Branches

During torpid pruning eliminate all fire blight affected branches on pear and apple trees. These blisters can be recognized by their dead, darkened, and indented appearance. Cursed terminals are darkened furthermore, frequently hold the dead leaves all through the winter. Starting soon after sprout, investigate trees week after week until midsummer for shoots and prods that might be tainted with scourge. Speedily eliminate the ailing divide 4 to 6 creeps beneath the clearly infected part if such a training is financially practical. Water sprouts shoot development, and spikes on the storage compartment also, base of the platform parts of pear trees ought to be taken out before sprouting. This will kill plant diseases and bringdown the focus to the terminal closures of these spikes and shoot development, hence forestalling scourge improvement that will regularly support the trees.

Peach and Nectarine Trees

Figure 7.4. Pruning Peach and other Nectarine Trees

Source: https://www.google.com/imgres?imgurl=https%3A%2F%2Fi.pinimg.com%2F600x315%2Fa6%2F01%2F1a%2

Peach and nectarine trees are pruned the equivalent way. Yearly pruning of bearing trees is basic to the creation of fruiting wood and significant returns of fruit products. Since the blossoms of these trees are found just horizontally on l-year-old shoots, all the fruiting wood should be basically recharged each year. Appropriate pruning will help the development of good fruiting wood. This turns out to be especially significant as the trees become more established. Pruning the bearing tree should likewise be done to help in irritation control and to accommodate positive light conditions all through the tree. Routine pruning can improve plantation productivity and the amount of attractive organic products created (Riccardo Gucci & Claudio Cantini, 2000).

Season of Pruning Peach and Nectarine Trees

Pruning of peach and nectarine trees should be deferred until the threat of a serious winter is past. By and large, pruning should be done from mid-March to early April. It ought not, in ordinary seasons, be postponed for such a long time that buds may be knocked off in the pruning activity. In the event that extreme winter temperatures have been capable, postponed pruning might be particularly useful. Postponing pruning implies the expulsion of buds that have endure can be dodged. Winter harmed blossom buds can be distinguished in pre-spring or late-winter by the trademark dull focuses when cut in the cross area. Wood that has been harmed by low temperatures will seem wrinkled or wilted furthermore when cut will show an earthy colored layer of tissue under the bark. Such wood can be effortlessly distinguished and eliminated by late-winter. In the event that low temperature injury to the wood was broad, it is frequently better not to prune the trees in the spring and stand by until summer to eliminate the dead wood.

Pruning Procedures

Peach and nectarine trees that have reached the bearing age require yearly heading back of old fruiting wood to sidelong shoots created on this wood, and cautious dispersing. Also, all harmed or ailing wood should be eliminated, just as that developing an offensive way, so eliminate holders, branches that become descending, and branches that meddle with the development of wood through the plantation by going to a parallel shoot or by dispersing, if 1-year-old shoots are accessible. In general, trees should be pruned to the degree that organic product can be reaped helpfully from a 6-foot stepping stool. When dispersing shoots, appropriate cuts over the tree so as not to leave a group of twigs at the closures of branches. The "horse tail" impact can be kept away from when pruning continues from the tip to the base of each framework branch. When choosing fruiting wood to save, those shoots that made 10 to 18 parts during development the past season by and large are the most gainful. It is most bountiful after serious pruning or loss of enormous branches. Such development may reach 5 or 6 feet and may present an extraordinary issue. Most water sprouts need to be dispersed during lethargic pruning, however a few may stay for forming into new platforms. These are made a beeline for an outward parallel shoot with some dispersing of the leftover laterals.

So dealt with, they may form into fruiting wood supplanting what was lost or pruned away. To keep up the trees at the ideal stature, it is imperative to head the upward-developing branches to solid, outward-developing laterals. It is desirable to make the cuts effortlessly by using hacking shears or short shaft pruners when remaining on the ground. The restoration point is first settled by heading back every fundamental branch when it comes to the ideal tallness. This cut is normally made in 1-year-old wood. In progressive years the restoration cuts should be made in the region of the first cuts; cut above it a few a long time and beneath it a few years. The focal point of the tree is maintained sensibly open in control to keep up the spreading structure (Riccardo Gucci & Claudio Cantini, 2000).

Abstain from moving in peach trees while pruning, particularly with hard-soled shoes. The bark can be handily scraped, which brings about open injuries where blister diseases can occur. On the off chance that trees are excessively tall to be pruned from the beginning, is ideal to prune from stepping stools. Extreme heading once again into more seasoned wood or "dehorning," might be worth some of the time. This training most coherently finds a way into the pruning program after trees have been permitted to become fairly tall from the absence of pruning whenever ice harm happened and crops were borne just in the upper pieces of the trees.

Chapter 8

Integrated Water and Nutrient Management of Fruit Crops

CONTENTS

8.1 INTRODUCTION

Food security is under attack and on the verge of collapsing. With an estimated 7.8 billion people under this marvelous universe, food demand has continued to spike while the expected source of survival continues to plunge. With the rise in technology and innovation, food demand has relatively superseded the world's expectations. Research by the Harvard business review group has it that by the year 2050, there would be a 98% demand for food and fruit products. This is in line with the expected rise in population to a tenth of a trillion in the near future.

However, fruits are lesser considered in our generations because of the constant rise in technology and children's upbringing. The resulting dietary changes that attract the consumption of proteins, snacks and meat products are alarming. Doctors have it that excessive consumption of such foods and diets may be detrimental to the health of many people. The age factor has drastically dropped with people roughly hitting the 60-year mark. Experts have drawn their attention in this seeking for individuals to consume more vitamins and agricultural products rather than chemically-made food. This is the importance of fruits in our generation as it is a guarantor of human life and an important aspect of the age factor (Valpuesta, 2002).

Fruit crops are generally produced by the agricultural sector, this incorporates the use of land, conservatories, arboretums and greenhouses. Unfortunately, this is the most hit platform of our generation. It is under attack by the alarming climatic changes and human demeanor. As a matter of fact, lesser individuals on our planet are concerned by how fruit crops are managed as they are not aware of the importance of fruit crops. It is estimated that around 10000 years ago, when humans began putting up structures, their environment was filled with green lush vegetation and crops. In an age and era that never knew of chemistry or industry, individuals believed and relied on natural land produce which attributed to fruit products, seeds and vegetables. It was the world that we would have never dreamt of.

Today, in an estimated space of a single human lifetime, seasons have changed. Nature and the availability of fruit crops have lost their stability. In this situation, the world has run out of fruit production, the essence and basic pillar of health and body nourishment. In an effort to recover and meet the current human demand, there exists a need to reignite environmental conservation. This proposes that we use media channels and means that do not pose a danger to the ecosystem

Figure 8.1: Fruits produced in a home garden. Due to rising populations, there is minimal land availability hence people apply the use of home gardens

Source; morning chores

One important aspect of the modern society is the proper use of the available resources. Water is a rare commodity in the current world. An estimation by the United Nations has it that more than half of the world's population in 2050 will be affected by insufficient water. Pollution and increased demand for water and water products both for consumption and agricultural use lie as the highest risk. This brings around the integrated water use in fruit farming. Water integration lies in a wider field known as Integrated Water Resources Management (IWRM). This field is entitled to ensure proper coordination and management of water sources and their use. It aims at maximizing the economic impact on matters of fruit production while reducing the vulnerability of its sources (Valpuesta, 2002).

Another important factor in fruit crop farming is the use of nutrients incorporated by fertilizers. Fertilizers are made of chemical compounds that are deemed very important in their handling. This implies that there ought to be a management factor relevant to its use. Having in mind that chemical compounds contribute to 80% of the destruction of water sources, nutrient management should be handled equally to water management. Inorganic fertilizers are the largest threat to the environment as they result in water contamination. Inorganic fertilizers are generally acidic and easily stunt plant growth

They are known to provide useful nutrients in plants but can have worse side effects on the fruit crop in future cultivations and on the ecosystem at large. Chemical engineers have it that the long-term use of chemical fertilizers and synthetic fertilizers easily lead to chemical imbalances in the fruit crops. During the early periods, one may get to witness good fruity production and maximum produce. This, under the influence of synthetic fertilizers, will easily degrade with time as the produce will get on reducing.

Chemical fertilizers are a key source of important nutrients such as phosphorous, nitrogen and potassium. They are generally very cheap as compared to their organic counterparts. However, one wise man had it that while cheap is expensive, expensive will always turn out to be cheap. In an event where a farmer would go on purchasing chemical fertilizers, they will have to brace themselves for tougher times ahead in as much as fruit production is related (Valpuesta, 2002).

Taking an insight of the nutrients found in inorganic fertilizers, ammonia nitrogen is known to be the key source of eutrophication. It further leads to ground water pollution which in turn leads to fruit poisoning. The high acidic content of the nutrients in inorganic fertilizers raise the soil pH, putting at risk the microbial ecosystem (this includes microorganisms such as the actinomycetes, protozoa, bacteria, fungi and algae). The microbial ecosystem is very important in that it ensures the cycling of other nutrients available and fed into the soil. Regular use of inorganic fertilizers eventually leads to a chemical imbalance in the plants obtaining nutrients. With this, the soil in question further avails itself to toxic buildup, creating a home for dangerous chemicals such as uranium, arsenic and cadmium. If such elements are availed to fruits, the fruits become poisonous and result in dangerous issues of human health. Nutrients apply to fruit crop farming as a double edge sword whereby productivity can be hiked at the expense of destruction (Hancock, 2008).

Below are areas that need to be focused upon to ensure a reliable water management system and the proper use of nutrients in fruit farming and production.

Figure 8.2: The application of inorganic fertilizers. If not managed, they pose a danger to fruit growth

Source; glocal khabar

8.2 SOIL NUTRIENT BALANCE

Soil nutrient balance is defined as the measure of the difference between the amount of nutrients entering the soil i.e. organic and inorganic fertilizers being placed into the soil and the nutrients leaving the soil i.e. the nutrients being used up by the fruit crops in the field. Maintain soil nutrient balance aims at ensuring that there is no nutrient depletion and on the other hand that there are no excessive amounts of nutrients in the soil. This implies that one inputs a nutrient content that is deemed necessary and that can be absorbed directly by the plants without much left in the soil.

Maintain soil nutrient balance requires that a farmer heeds to two different stages. First and foremost, the farmer in a fruit farm is entitled to diagnose their respective farm. This is the diagnostic phase. This is an analysis phase that is farm-based and requires that the farmer draws information from the farm. It incorporates the quantification of different nutrient levels in the farm. It further quantifies the relative flow of nutrients between crops, animals and the soil in relevance to the type of fruit crop that is to be planted. Flows in the nutrient perspective can be analyzed in terms of respective kilograms of different types of nutrients or a financial perspective (Hancock, 2008).

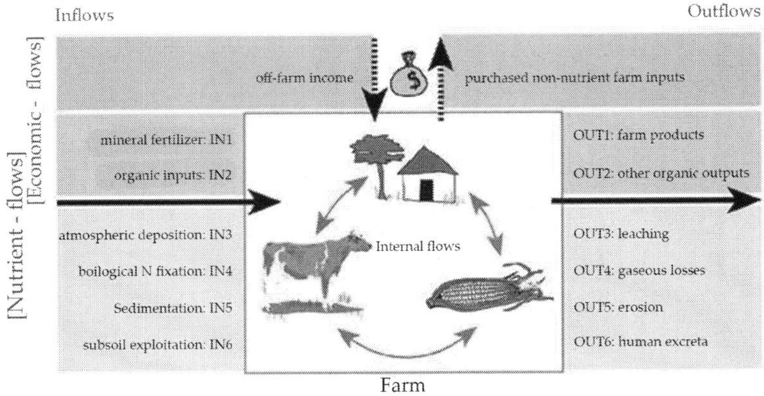

Figure 8.3: A figure illustrating the concept of soil nutrient balance.

Source: microlevel

Quantified nutrient flows entail an explanation of the kind of activities that are either consuming nutrients or accumulating nutrients for the farm. It further tells of how nutrients would flow from one particular point to another.

Financially, quantitative flows tell of the profitability of activities that relate to the accumulation or consumption of various types of nutrients. It also brings forth the labor demands of the farm in relation to the same activities. The diagnostic phase also entails the farm's history, its initial problems relating to nutrients and possible solutions as detected by the farmer at the said period of time. After this, the farmer now enters the development phase which incorporates the use of machinery and technology as per our current society (Zhang, 2017).

Technology desires procedures. The first stage in this is to check the analysis brought about by the diagnostic phase. With this, the fruit farmer can come up with ways of identifying and developing various kinds and types of technologies to address the given imbalance or rather types of problems. For example, a farmer may find out a nutrient imbalance whereby the soils reflect a negative nutrient balance. This may be caused by huge volumes of erosion and leaching. To deal with this, modern means of preventing erosion are applied with the aid of technology e.g. the use of tractors to build terraces and conservation tillage. The application of technology in this phase also helps to provide quantitative and scientific information which is very important when dealing with nutrients. The use of computers and the internet when conducting research is an example of this. In the development phase, there ought to be a correlation between science-based knowledge and local farmer knowledge. This phase brings in the different types of techniques, both traditional and modern to maintain soil nutrient balance (Zhang, 2017).

The application of cover crops

Cover crops in this case are the plants that will be used primarily to maintain nutrient balance. They can be applied in two situations. First is whereby there is a negative plunge in the availability of a particular nutrient. The cover crop will hereby be one that retains the valuable nutrient. In the other case, the soil may be having excessive amounts of a particular nutrient. The cover crops will come into action by removing or absorbing the excess nutrients. For example, a farm may be having the nutrient sodium in excess. The farmer will be advised to grow much of the mustard fruit to suck up the high sodium content.

When the fruit-plant is ready for transplanting the mustard can be removed and taken as compost to a sodium-deficient farm. The use of cover crops is very advantageous in that one can interplant crops that are both very productive. This will instead maximizer profits in favor of the lost portion of land.

Figure 8.4: The use of cover crops. A Nebraska farm intercropping maize with vegetables as cover crops.

Source; University of Nebraska

Mulching

Mulching is the effective use of plants or plant materials/parts by spreading them on the farm floor surface. More than a means of preventing soil erosion, mulching goes a long way in providing important plant nutrients for the fruit farm. Different forms of mulch take different ways for them to break down. For example, when using softwood or sawdust as mulches, they are generally in constant need of nitrogen for the break-down process. In this case, the mulch will absorb nitrogen from the soil before it commences decomposition.

A farmer may propose removing the mulch so that the nutrients are removed. This is in an event where the nutrient i.e. nitrogen was in excess. On the contrary opinion, another farmer may leave the mulch to decompose hence returning the nutrients back into the soil. However, this will in time increase the amount of nitroge4n in the soil. One may propose transferring the mulch to a nitrogen-deficient fruit farm.

Leaching

Leaching can be defined as the extraction or loss of nutrients from the soil by a carrier normally a solvent such as water. This can be done naturally through the effect of rain or intentionally through the irrigation media. Since it only entails the removal of nutrients, leaching is carried out on farms that have a certain nutrient in excess. It can also be hazardous in that it attributes to the loss of nitrogen in many farms. In arid regions or areas on the verge of desertification, sodium is known to be in excess. Such areas are known for the growth of mangoes.

Sodium is commonly identified for such reasons as a similarity to salt. A farmer would be advised to mix sulfur with the soil or add gypsum to the soil. After few weeks, the farmer can run water over the soil to kick off the process of leaching. Arid conditions usually do not favor the growth of fruits. However, farmers can get to use the availability of short sporadic rains to ensure proper leaching.

Rotational grazing

Rotational grazing involves the use of livestock in controlling the nutrient level in farms. Many people know of the lawnmower, the famous machine that cuts down grass. Then there is the natural lawnmower: ruminants. Ruminants are the kind of herbivorous animals that graze in the fields. They include cattle, sheep, and goats. They are great providers of organic fertilizers and are loved by most fruit farmers. The good thing about ruminants is that they can deposit organic manure whenever and wherever you need them to.

A farmer may decide to place the ruminants in a farm field where they can graze and add a certain kind of nutrients to the area. However, to ensure that the nutrients are not deposited in excess, the farmer ought to transfer the animals to a new field so that they may not keep filling the same portion of land with the same types of nutrients.

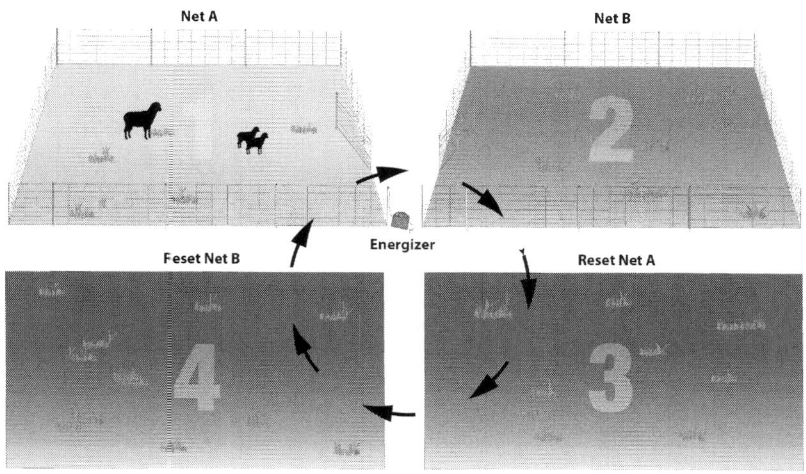

Figure 8.5: An image illustrating rotational grazing

Source; premier 1 supplies

Crop selection

Crop selection intends to cover the interest of the soil rather than the interest of the farmer. This comes up in a situation whereby a fruit crop is planted in relevance to the kind of nutrients that is in excess in a particular track of land. For example, a piece of farmland may bear nitrogen in excess. Rather than coming up with ways of removing the excess nutrients, one may be advised to plant a crop that is in relevance with the nutrient or one that needs the nitrogen nutrient. To create a source of profit for the farmer, the crop will turn the nutrients into important food for the farmer.

In an event where the farmer happens to have insufficient nutrients in his/her farm, the farmer may sow daikon radishes which will hold up and boost the nutrients till the next planting season.

This ensures that the soil is rebalanced to its original state.

- **Balancing with compost:** The use of compost is one of the best means aimed at ensuring nutrient balance in the soil. Compost manure is organic and relatively less hazardous to the soil. It is considered the only safe additive to the soil during planting. It further helps to reduce various fruit diseases. It has massive abilities in balancing the levels of soil ph. and build soil structure.
- **Proper organic fertilizer rationing:** This is the most modern way of ensuring a soil-nutrient balance. This involves the use of technology to test for soil inadequacies. Experts always advise the application of chemical fertilizers to cover up for any form of deficiency in the soil. One known organic fertilizer used in farms is the NPK. This represents Nitrogen, Phosphorous and Potassium. This form of organic fertilizer comes with different levels of content for the said nutrient.

There is however a natural remedy for this entailing the use of either feather or bone meal in the need of nitrogen, rock phosphate for phosphorous and wood ash and greens and for potassium.

8.3 THE PLANT NUTRIENT BALANCE SYSTEM

Nutrient balance in the soil or in any other medium that is used to grow fruits is very essential and this helps to provide information about the pressures in the environment. When there is a nutrient deficit or the value in the nutrient amount in the soil becomes negative it shows us that there is a decline in soil fertility. On the other side when there is nutrient surplus i.e. positive values then this show the risk of polluting the soil, air and/or water. Nutrient balance can therefore be described as the difference between the nutrients that are added to the soil and these may include livestock, green or farmyard manure also referred to as organic fertilizers and fertilizers commonly known as inorganic fertilizers and the nutrients that are leaving the soil system and into the crop (T. Radha & Lila Mathew, 2007).

When there is excessive nutrients in the soil which are in excess of what is needed to the crop it leads to nutrient losses which may lead to an economic loss to the farmer and/or even potential harm to the environment which include ammonia and greenhouse gas emission. In order to understand the physiological requirements in trees will include the study of seasonal variation in leaf nutrient contents as this is essential in determining fertilizer estimation in fruit orchards. The link between the period and the rate at which fertilizer is applied within the plants' demand for the nutrients may help to maximize the yields and also help to increase efficiency in nutrient use. In trees like the *Citrus reticulata* Blanco which is an example of mandarin trees, the time when they fruit there is a significant effect on the potassium (K), phosphorus (P) and magnesium (Mg) in the leaves and the stem, the reason being the fruits act as a strong sink for potassium and phosphorus; during the harvesting stage, the amounts of K and P decreased to a deficient level (FAO, 2020).

Figure 8.6: A *Citrus reticulata* orchard. It is an example of mandarin fruit trees.

Source; researchgate

During the growth of perennial fruit crops, there is a specific demand for a targeted nutrient. The accumulation of nutrients is thus considered to be a continuous process as the tree is growing and the nutrient that is needed is determined by the stage of growth. The fruit can absorb nutrients such as calcium (Ca) as early developmental stages and as it grows then it absorbs the other macronutrients. In many instances, it has been noted that boron(B) has a significant effect during fruit setting which means that when there is a deficiency of boron during flowering and fruiting then this leads to fruit abscission and a reduction in yield and the quality of fruits becomes poor. Many studies have shown that there are nutrient dynamics in fruit trees, especially during the growing stages. Deep analysis has to be done on the physiological aspects of tree nutrition and it will depend on the knowledge of the seasonal variation of the content of nutrients in the leaf. These factors tend to be dynamic throughout the year hence the information obtained from the analysis will provide the appropriate recommendation of fertilizer amendment according to the stage of growth of the plant. This will help the farmer determine the seasonality of some macronutrients such as; nitrogen(N), phosphorus(P), Potassium(K), Calcium(Ca), Magnesium(Mg), and some micronutrients such as; Copper(Cu), Manganese(Mn), Zinc(Zn), Iron(Fe) and Molybdenum(Mo) in their fruit orchards and the relationship between the nutrients and the plan phonological stages. In tropical fruits, it was noted that N remobilization was very high in older parts as compared to new parts and from absorption from the soil (FAO, 2020).

Figure 8.7 The Satsuma mandarin. Shows seasonal variation in nutrient requirement.

Source; Struble Ranch

When there is excess P in the soil then this leads to inhibition of uptake of other nutrients. This shows that soils with less organic matter and receive high amounts of P, the soils may exhibit excess Zn levels because these two are proportional in soil. Iron can also be deficient in soils especially when there is heavy rainfall. The amount of Mg is greatly affected by the phenological stage of the fruit for instance during flowering and fruit set. The amount of Mg in the fruit trees could be low due to the fact that Mg is involved in numerous enzymatic reactions that involve the provision of energy to the tree (FAO, 2020).

In many temperate fruit trees such as peach, apples, grapes , persimmon and *Satsuma mandarin* show a seasonal variation in their nutrients, especially nitrogen (N) and phosphorus (P). The concentration of P and K were low during fruit formation, this shows that P and K could be sinks that are involved during flowering. This means that during fertilizer application these two elements or nutrients which are mostly included in the fertilizers are essential and will affect the yield and quality of fruits. Mandarin trees have shown the best nutrient balance and a fruit tree such as Persimmon has the most unbalanced in terms of mineral nutrition. It is important to note that when there is improper management the continuous production of crops will reduce the reserves of nutrients in the soil and the sustainability of soil will also be affected due to the fact that people want higher yields. A fruit farmer should always ensure that their approach to farming is eco-friendly and will maintain the ecological balance of the soil ecosystem.

8.4 PLANT WATER BALANCE SYSTEM

Water is very essential in all plants as it is involved in several processes. Water can be fed to plants through two main ways i.e. rain-fed or through irrigation and sometimes one can include both. Fruit trees need water as it is involved in the production of the wood structure, fruitfulness of the tree, sizing of the fruit and also the cooling of the canopy on warm days. In some plants such as peaches, water accounts for almost 70% of its final volume during the last 30 days of the fruit on the tree. As the trees are growing and they are grown in dry soil conditions this will inhibit canopy formation and this leads to delays in fruit formation. In the case of apples when they get less water or inadequate rainfall it will lead to a bitter pit (the fruit will have black spots) (Robert E. Paull & Odilo Duarte, 2011).

Figure 8.8 Bitter pit in apples. Caused by lack of water and plants grown in dry soils.

Source; twitter

8.5 CONSTRAINTS IN USE OF ORGANICS COMPLEMENTARY WITH MINERAL FERTILIZER

Fertilizers are used in everyday life to supply nutrients to plants. Most farmers prefer the use of fertilizer in supplementing their crops even though they are considered to be quite expensive. Given their expensive nature, the assumption is that most farmers would decide to transfer or rather move to the adoption of organic manure. Their popularity among most farmers including fruit-producing farmers is due to the fact that they are easily used with minimal cost used to transport, with little labor and the time used is always minimal. Basically, the contents in the fertilizer are considered to be concentrated amounts of plant nutrients. Such nutrients are then manufactured to come up with convenient packs of fertilizers. There exist different kinds of fertilizes including those that are applied just prior to planting, those applied during the planting period and those that are applied sometime after planting. The use of fertilizer is quite popular especially in the urban and peri-urban areas where the level of awareness is conveniently high. Consequently, most of the time when fertilizer is used is when a farmer is planting for the purpose of commercialization and the crops that are irrigated in most cases. Fertilizers are normally made for the purpose of ensuring that the producer's crop yields are high and therefore the reason as to why it has gained widespread acclaim among the commercial or rather the market-oriented farmers. Such farmers know that there is a need for high and better production in order to raise enough revenue to run their day-to-day activities especially in areas where crops need to be irrigated because of the various weather constraints (Robert E. Paull & Odilo Duarte, 2011).

It should be noted that for a producer to have a bumper harvest, the soil that is used in the planting of his crops needs to be healthy. To some extent, nature is always trying to replenish the soil to make it suitable for plants to grow. However, with the always-changing environment and the difficult climatic conditions, it is quite a difficult task. As such, nature requires help in the replenishing of the soils and thus the development of fertilizers that are suitable for that job. When the producer carefully analyzes the crops that he is planting and the fertilizer used, then there will be an ability of the chain between humans and plants to be completed, in which the plants will be able to provide humans with nutritional food. The chain that is completed involves the fact that the fertilizers feed the soil, the soil then feeds the plants while the plants have the ability to provide humans with nutritional products. The most common nutritious fertilizers among producers include phosphorous, potassium and nitrogen. It is argued by most scientists that plants have a limited ability to avoid plant nutrient deficiencies if they are not provided with a little bit of assistance by the fertilizer. Nitrogen is found in the air but it proves to be quite difficult for plants to absorb it. As such, absorbing it from the soil is the only way for them to acquire this crucial nutrient. If for instance the soil has been used to plan a number of times then it will be nutrient deficient, as such, nitrogen is required for nutritional levels of the soil to be boosted. For the case of potassium, they are always very deep from the surface of the soil. As such, plant roots always find it difficult to access them and thus the need for fertilizer. Furthermore, through the mining of potassium, it can be said that it helps bring this important nutrient to the surface of the soil (FAO, 2020).

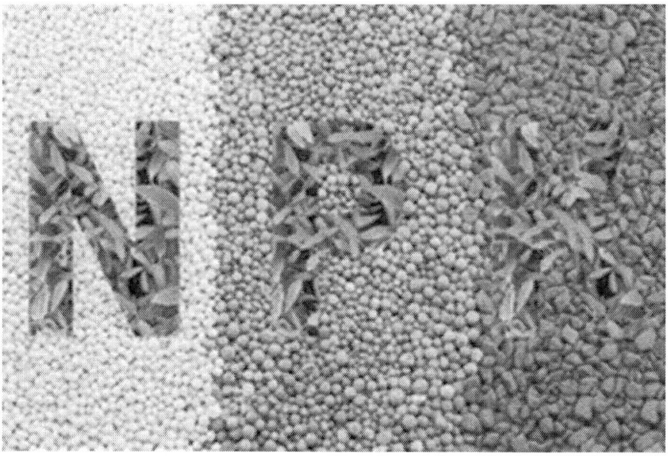

Figure 8.9: The most common mineral fertilizers among farmers are Nitrogen Phosphorous and potassium.

Source: IndiaMart

Phosphorous on the other hand is quite important in ensuring that there is high production and also ensuring that the plant growth is rapid. These minerals are common in certain types of rocks. For plants on the other hand to acquire this mineral, it requires it to be diluted in water. In the right way, nature requires that this mineral is absorbed through the soil. All these nutrients, phosphorous, nitrogen and potassium are ideally important to the plant and the right way in which plants can absorb them is through the soil. There are actually more than 15 nutrients in the world and the type of nutrient to be used depends mainly on the type of plant the farmer wants to grow and further, he should consider the type of nutrient that is lacking in the soil.

Organic manure is also important to the plants and quite frankly, it are easily acquired, cheap and highly reliable to ensure that plants are able to grow. However, in most cases, scientists do not allow the use of these nutrients concurrently. It is either this or that but not both as it is considered to be too much for the soil. It is true that for centuries, organic fertilizer was being used as an important source of nutrients and rather organic matter and was considered to enhance the structure of the soil. Such organic fertilizers include compost manure from plant waste and animal waste. The introduction of mineral fertilizer was because of the growing human population in the world. This meant many mouths to feed and therefore, need for a way to increase the production of food to ensure sustenance. With mineral fertilizers, most producers can only use less amounts as compared to when using organic fertilizer. The use of a certain type of fertilizer as seen depends entirely on the type of crop that is planted and the type of soil the plant is being planted on. For the case of mineral fertilizer, most farmers tend to use them whenever they are planting grain products such as maize wheat, rice among others. Commercial crops such as sugarcane, tea and cocoa, cotton and bananas tend use mineral fertilizer as well. For the case of vegetables and fruits, the farmers tend use organic or rather compost manure. In most instances when farmers decide to use green manuring, it is mainly common in sugarcane and rice. In the case of sub-Saharan Africa where the climate is arid and semi-arid, there is a prominent use of farmyard manure. This is due to the fact that; mineral fertilizer is discouraged as it is assumed that the use of such fertilizers will be harmful to the soil given the climatic condition of the area. Furthermore, in these areas, most of the farming is done by irrigating the plants and thus that much water will render the fertilizers useless to the plant. In such areas also, the use of farmyard manure is considered to be very important because of its ability to store moisture for a longer period.

Figure 8.10: Some of the popular fruits are grown in the arid and semi-arid areas and thus the only way of providing them with nutrients involves the use of organic manure.

Source: International Tropical Fruits Network

Fruits are some plants that are considered to thrive in arid and semi-arid areas. As such these farmers prefer the use of organic manure in the provision of nutrients for their plants. For instance, in India, most fruit farmers use organic manure, including compost, farmyard manure, goat, poultry and pig manure among other animal waste. This is because, by use of organic manure, the soil in those areas will have the ability to regenerate the fertility of soil after some few years of cropping, which maybe three to five years. There are various other constraints apart from climatic conditions that prevent farmers from using a certain type of fertilizer. One of the constraints involves the agro-ecological difference of different areas (NIIR Board, 2005).

In arid and semi-arid areas as seen, the moisture content is quite low and the farming practices that are used involve the use of irrigation. Such areas are discouraged from the use of mineral fertilizer as they are quite harmful to the soil. As such, these areas tend to prefer the use of organic manure due to their ability to retain moisture content for longer periods. Furthermore, their corrosive power is quite limited as compared to mineral fertilizers. Another constraining factor involves the number of enterprises a farmer has on his farm. For instance, if a farmer has only a single enterprise, that is only does farming and does not rear cattle, such farmers are always forced to depend entirely on the use of mineral fertilizer and moderately use of plant manure. As such. The farmers are constrained to the use of organic manure because of their inability to access these products. On the other hand, for the case of farmers that practice farming and also rare animals such as cows, pigs, goats, and sheep among others have a higher probability to practice organic farming or rather use organic manure in many cases as compared to the use of mineral fertilizer. These farmers always have the ability to recycle the animal waste and use them on their farms and at the same time providing the farmers with profits and further preventing them from depending on the costly fertilizer inputs (Valpuesta, 2002).

Another constraint that farmers face in the type of fertilizer to be used is land tenancy. For the case of farmers who farm on their own land, most farming instances involve the use of farm yard manure. This is because these farmers have the ability to decide whether to rare animals or not. Furthermore, most of the products that are planted in their farms are for subsistence use only and for those who have decided to commercialize only do it partly. However, for the case of those individuals that hire land, most of their practices tend to be for commercial purposes. As such, it is only human that these individuals choose to use mineral fertilizer instead of organic manure. Their attributes of being portable and highly efficient when it comes to production make these products perfect for such individuals. The use of mineral fertilizer has been widely praised by farmers as it ensures high productivity and therefore increase in profits. Such profits are considered to be important as they have the ability to cover the cost of renting the land in the first place. Another key thing is that these farmers that rent land for the purpose of farming always practice cereal farming. Mineral fertilizer is quite efficient as seen above in planting or rather farming cereals.

Figure 8.11: Whenever land is leased, most individuals prefer the use of mineral fertilizer.

Source: Oxfarm

One discovery that has become very constraining to plants when it comes to the efficient use of organic resources is the fact that these resources are having more and more competitive uses. For example, the use of cow dung as fuel has been praised by most individuals as it is considered to be more preservative to the environment. Plant stalks such as those of castor, cotton and red gram are also used as fuel. Furthermore, over the years, fuel availability has been more and more constraining because of the various measures that have been put in place by various governments denying members of the public from cutting down trees. This is because there has been a lot of distractions to the environment and thus trees are currently considered to be an important resource for the environment. Apart from using these products for the purpose of fuel, some of the plant residues are commonly used in the manufacture of animal feeds. Also, some of the animal droppings such as those of poultry are used as additives when manufacturing some of the fish and cattle feeds. The fact that these organic resources are seen to be more and more useful, some of the traders that deal in the supply of these products are using this to their advantage by raising their cost (Valpuesta, 2002).

In the peri-urban areas, organic manures are highly demanded since individuals have ornamental gardens where they plant vegetables and a bit of fruit trees. The high cost of such products has thus forced the majority of the farmers to turn to the use of mineral fertilizer and little organic manure. Furthermore, the cost of transportation is quite high making their purchasing cost to skyrocket. Organic manures are quite bulky making them difficult to transport and therefore difficult for farmers to use them in all seasons. As such, it is only easier for farmers or rather these traders to transport small quantities of the same. As such, the application of organic manure is only conveniently used in a sufficient manner in certain crops. Such crops need to be remunerative in a way and can only be used within an interval of 4-5 years and not beyond those years. This is especially done to those crops that are considered to be *kharif* crops. These are crops that are grown during the monsoon season and are domesticated by the producers of such crops. They include crops such as rice. This is done in areas such as India, Pakistan and some areas of Bangladesh.

Figure 8.12: Organic manure is quite bulky and even though efficient, farmers prefer the use of mineral fertilizer.

Source: Amazon.in

Pests, diseases and weeds are one constraining factor to producers that prefer using organic manure. Most farmers believe that the use of organic manure can be very destructive to their plants. This is because they believe that, organic manure carries pests and diseases. Most of the small insects prefer the warmth and the comfort of making animal and plant waste their habitat. As such, when they are used to provide nutrients to the plants, these pests climb up the plant and become very destructive to the produce. For plant residues that are used as organic manure, if they had any form of disease, then they can easily transfer it to the produce and thus a huge loss for the farmer. In some cases, when farmers are tilling their land, they decide to use some of the weeds for the purpose of mulching. However, there are some weeds that are considered to be very stubborn and thus do not dry. If these weeds are mixed up with other weeds and used for mulching, they can be able to develop roots and grow at a particular point. If this happens, then the weed will be taking up the nutrients of the said plant and therefore leading to them being unhealthy. For the case of fruits, when the weeds take up the plant nutrients of the said plant, then the fruits that will be produced will be quite small and in worst-case scenario, the fruits might fall on the ground before they are ripe or are rather ready to be harvested (Nandwani, 2014).

As stated, it is seen that organic manure as well as mineral fertilizer are very important to different farmers. Also, regardless of where a particular person is or the condition of the soil, there will always be an alternative for them to use, either organic or inorganic manure. Seeing such importance of organic manure, researchers have done countless researches with regard to integrated nutrient management. This has especially been done in natural management institutions and the various agricultural universities that are owned by the state. As such, farmers have adopted various Integrated Nutrient Management practices for some of the crucial crops all over the world. Some of these crops are mainly fruits. This is due to the fact that most countries are advising their citizens to increase their consumption of fruits and vegetables as they provide major nutrients to the body. As such, producers have had the luxury of knowing and understanding the role of the various organic manures and how they are becoming useful in increasing input efficiency and how much they are considered to be favorable to the physical and biological conditions of the soil. Furthermore, there has been an improvement in the know-how with regards to how important a role integrated use of organic manure plays in ensuring that nutrient cycling in the plant system is efficient quite frankly in different production systems and further in different types of soils. It has also been pointed out that organic manure helps in mending the physical structure and the biological structure of soil and thus contributing to the research on Integrated Nutrient Management. Several institutions have been able to work hard and have had the ability to publish books and journals with regard to the same matter.

The promotion of this project however is seen to be facing quite a huge challenge. This is especially in areas that are rural where literate levels of individuals are quite low and change is feared. As such, it will require that the actors that will be involved put more effort in the promotion and further, the project will require a number of actors and more so a large number. As a matter of fact, for there to be a success in the implementation of integrated nutrient management, the various research institutes need to do a baseline survey by implementing the project in a single area and observing how it will turn out to be. Furthermore, they need to look at the participation of individuals in the project. For there to be the successive implementation of integrated nutrient management, a good number of farmers need to be willing to involve themselves in the project. Furthermore, if one wants to know if a certain project will be successful, one looks at how much some of the major organizations such as the NGO, the Government and the private sectors are willing to invest in the same. If such institutions show an interest in investing in the same, then it shows that the probability of success is quite high (Nandwani, 2014).

One of the major strategies in ensuring that the integrated nutrient management project is a success is by educating farmers on the importance of recycling crop residue and green manuring. Ideally, it does not mean that if one does not rare cattle then he/she will not be able to practice organic farming. Management of the crop residues can be done in various ways including burning, removal or rather deciding to incorporate them into the soil. Burning however, is considered to not be efficient as in such situations, the soil nutrients can easily be lost and further, it leads to the pollution of the environment. Instead, individuals are advised to use their residue in the creation of bio-fertilizers. These products have a high Nitrogen fixation rate making them efficient in adding nutrients to the soil. Another crucial role that is played by these bio-fertilizers is the fact that they help in liberating growth substances. These substances are crucial in germination and ensuring that there is a growth of plants.

It is seen that organic manure is quite important to plants and ensures that the research on integrated nutrient management on plants is effectively done as it will helping ensuring the increased production of fruits, enough to sustain the world (FAO, 2020).

8.6 BETTER WATERING SYSTEM (IRRIGATION).

Irrigation is one of the most sought after ways of farming in the world. With the ravaging effects of desertification, there is need for the world to embrace irrigation. However, to ensure an integrated means of water conservation, there need to be better watering system which will instead ensure that less water is lost during the irrigation process. For example, the application of drip irrigation as compared to other forms of irrigation helps save more water by reducing t5he amount of water left for evaporation. Rather than placing water droplets on every corner of the farm, drip irrigation refocuses on the base of the plant and ensures that the water is effectively awarded to the root.

Figure 8.13: Smart irrigation. Incorporating technology in irrigation

Source; ovio smart world

In situations whereby technology has to be used, the application of electric sensors is of more importance. Instead of letting the water flow to the fruit farm each and every moment, it would work better and save more water if a sensor would let out water at different timings with this, more water is saved and can be put to use later on. It is also advisable for farmers to irrigate their farms in the morning or in the evening. This is a way of evading evapotranspiration. Evapotranspiration is the evaporation of water from plants through the leaves and from the land. It is at its peak during the midday sun. It is also advisable for farmers to use renewable sources of energy e.g. solar power. This assures the farmer to freely apply the use of pumps without the fear of electricity cost (Zhang, 2017).

In an effort to reduce water wastage, fruit farmers should avoid hit and miss watering systems. The root should be the center of focus while the farmer struggles to avoid hitting the leaves.

- Mulching: Mulching is one of the ways of preventing water loss from the soil. When handling a fruit farm, water is very important in that many fruits are entirely dependent on water. If one aims at avoiding getting to irrigation every now and then, it is important that the farmer takes up crop parts and lays them o the farm. This prevents the soil from direct attack by the scorching sun. Mulch can include straws of plants, wood chips, saw dust, pine needles, shredded bark and chopped leaves. Mulching also helps in improving the health of the soil. Healthy soils work a great deal in retaining more water.
- Conservation tillage: This is a farming method whereby 30% of the soil is tilled leaving behind more crop residue. It implements the use of special plows and other farm machinery. This helps reduce evaporation while increasing water absorption. An example of the use of conservation tillage is the combined use of mixed mulch, compost and the application of cover crops with no tillage.

Chapter 9

The Use of Greenhouses For Fruit Crop Production

CONTENTS

9.1 INTRODUCTION

Different factors, like climatic conditions, water availability, and soil quality, affect plant cultivation. Methodologies have been applied to either adapt food crops to their climate (such as by breeding and selecting plants that are more resistant to drought or have shorter development cycles), or to adapt the environment (such as supply of nutrients and temperature) to fulfill the demand of plants. Techniques such as the use of windbreaks, mulches, plant, or row covers, and cold frames have been included in practical means of changing the climate surrounding the plants. These plant protection methods can be defined as passive methods since they only create obstacles and do not regulate the atmosphere between the plants and their environment (FAO, 2018). Some forms of garden frames can be heated by artificial means (hotbeds, heated frames) but do not necessarily allow for any environmental regulation. Greenhouse production is the only form of production of food crops that makes use of environmental control. Also known as controlled environmental agriculture, modern greenhouse production is popular.

Figure 9.1: Increased competition for Armenia's greenhouse products has provided farmers with fresh export opportunities, not just berries, but also vegetables and flowers. Armenia's greenhouse sector needed to modernize to satisfy export conditions, particularly in brand new markets in the Middle East and Europe, in addition to Russia.

Source:https://www.ifc.org/wps/wcm/connect/news_ext_content/ifc_external_corporate_site/news+and+events/news/impact-stories/armenia-investments-greenhouses-bear-fruit

It is possible to grow food-producing plants through the use of a greenhouse in areas and at times when climate conditions negatively impact them or even prohibit them from developing. Also, greenhouses may be used to protect the crops against weather phenomena (such as wind, heavy rain, or hail) that would adversely affect them when climate conditions allow outdoor plant production. For the point of this study, the word 'greenhouse' is defined as a framework for the cultivation of plants covered by a transparent or translucent material in which environmental factors can be changed or managed. Tunnels are often used but are not generally called greenhouses, to change the environmental conditions for the development of plants. Because the difference between greenhouses and tunnels may not always be evident in the literature, all structures will be considered together when they are high enough for individuals to travel and work freely in them. While greenhouses were in operation since 1800 (or earlier), and in the second half of the nineteenth century, greenhouse agricultural production began to evolve as an industry, after World War II, the greatest growth and development of the greenhouse industry occurred worldwide. Today, greenhouse agriculture can be located on all continents (Cees Sonneveld & Wim Voogt, 2009).

The modern greenhouse is typically a framed structure enclosed in glass or plastic that is utilized for the cultivation of fruits, vegetables, flowers, and any other plants requiring unique temperature conditions. The main fundamental types are the greenhouse span-type, which has a roof double-sloped, or A-shaped, and the greenhouse lean-to, which has only one slope of the roof and leans against the side of a building. Often two or more span-type greenhouses are joined side by side so that they have fewer exterior walls, and thus the cost of heating is less. On its sides and top, a greenhouse has a wide expanse of glazing so that the plants are exposed for most of the day to natural light. The traditional glazing material was glass, but plastic films are now popular, such as polyethylene or polyvinyl, and fiberglass.

The framework is framed with aluminum, galvanized steel, or woods like redwood, cedar, or cypress. The greenhouse is heated partly by the Sun's rays and partially by artificial means, like steam, hot water, or hot air that circulates. Some form of a ventilation system is often necessary because a greenhouse may become too hot as well as too cold; this typically consists of roof openings that can be controlled manually or automatically, and end-wall openings from which electric fans draw air and distribute it throughout the interior.

During the winter, the first greenhouses were established to preserve the orange, lemon, and fig trees. Such Victorian structures were heated to a minimum of 18 ° C (65 ° F). Therefore, a tropical greenhouse concept arose as people began to cultivate bananas, pineapples, and other exotic fruits. Growers today tend to use cold frames and unheated greenhouses for vegetables and ornamental plants. For a few exotic plants, running a heater during the year can be quite costly and temperatures for other plants are quite high. Nevertheless, you can also find a spot to grow some fruits in an 8 ft by 6 ft greenhouse. The best fruits to grow in a greenhouse are strawberry pots and melon bags. Generally, when determining what fruit to grow, the temperature and the size of the greenhouse are considered. For instance, while oranges thrive in a heated structure, figs, and apricots like cool conditions. Many growers prefer to keep fruit trees in pots because moving them around is easy. Apricots, peaches, and nectarines, however, are better grown on the soil (Cees Sonneveld & Wim Voogt, 2009).

Grown in greenhouses are fruits such as fig, cherry, mandarin, orange, pineapple, banana, strawberries, and grapes, as well as culinary and medicinal herbs. In a greenhouse, fruit tree cultivation is not as common as you might imagine. You can be limited to growing this outdoors by your local climate. But one can eat and harvest their very own tropical fruits using a greenhouse. For planting such fruit trees, the general setting in a greenhouse is much easier. Various requirements must or should be met, depending on the form of fruit, so that the tree flowers and produces fruit. This involves a controlled atmosphere and some degree of commitment.

While the treatment of fruit trees in a greenhouse is identical to any planted outside, some particularities need to be considered. It is not feasible to plant all species in a greenhouse, to begin with. There are fruit trees more fitting for greenhouse production, either due to size or other requirements, and those that do not do well. Everything coming from the Mediterranean, subtropical and tropical climates, unless it reaches a certain height or width, would do well in greenhouses.

To produce fruits and flowers, some fruit trees need a winter's chill. They go through a certain period and they will not be supported with this by a warm greenhouse. If you really want to grow these indoors, during the cold season you must use a cold greenhouse or transport them outside. If you choose the latter, you will have to grow fruit trees in big containers. Moving these large pots is, however, not an easy job. One can get a Dolly from Potwheelz. These make the work much easier.

Some of the reasons why one would prefer to use a greenhouse includes:

- Perhaps the region is getting too cold for any of the varieties needed to be grown
- One requires fruits sooner, and that can easily be done in a greenhouse
- Fruits tend to grow in greater and larger numbers
- Perhaps one would like to shield their cherries and other fruits from the cruel birds that take them all away.

9.2 COMMON TIPS FOR GROWING FRUITS IN A GREENHOUSE

- *The Size of the Greenhouse:* They would need a big greenhouse with a high ceiling, no matter what fruit tree a person is going to choose. An 8-8 with 7.5ft at its height is not going to be enough. More than just a tree, one would want to be able to expand and will need to hit certain plants. In certain instances, for pollination, one may even need two trees. As tall and wide as the budget requires, it should go. In the middle, the minimum height should be around 9ft. The farmer has some room for growth in this way.

Figure 9.2: Until you start constructing or buying a greenhouse, think about what plants you are going to fill it with. Plants have varying requirements for temperature and humidity, and some greenhouses can have more insulation than others.

Source: https://daisylinden.com/types-of-greenhouses/

- *Picking Suitable Fruit Trees:* The farmer must think about the ultimate strategy for their greenhouse before running into the shop. They have got to harmonize the plants they pick. 2-3 distinct climates cannot be achieved. If they grow tropical plants, they should stick to them and not mix them with temperate plants. If they just start the seedlings outdoors and transplant them. They will be perfect then.
- *Pollination:* Without pollination, growing fruits does not function. There are a few self-pollinating fruits, trees and many others that require insects or wind, or even cross-pollination. This means that users either must attract pollinating organisms into their greenhouse or manually pollinate themselves. At least 2 trees would be required for fruit trees that need cross-pollination (e.g., apples, pears). Effective pollination helps in the formation of fruits and seeds. Some of nature's pollinators include birds, bees, butterflies, and wind. Greenhouses build barriers, which gives protection from unfavorable environmental conditions but also create functional barriers which prevent pollination. A little human interference ensures that plants, even in safe greenhouses, can be pollinated effectively.

The different forms of pollination include:

- *Manual pollination:* If you have a small greenhouse, it is cost-effective to manually pollinate plants and takes little time. Gently moving plants or touching flowers transfers pollen into female structures from male flower components. Some plants have different female and male flowers, like squash, so it is important to pass pollen between the blossoms. Some plants have 'perfect flowers,' which have male and female sections, such as tomatoes. Disturbing flowers disperse pollen inside each flower on these plants.
- *Device pollination:* From greenhouse supply companies, you can purchase battery-operated pollinating instruments. There are vibrating heads on hand-held pollinator wands that you tap at the base of the flowers. You can also pollinate flowers using a battery-operated or electronic toothbrush. They normally do less harm to flowers and fruits since pollinating devices are made specifically for this purpose. You can pollinate plants between 10 a.m., whether you prefer manual or computer pollination, or 3 p.m.

- *Bee pollination:* Bees are, as nature intended, ideal pollinators. Bumblebees, due to their longer tongues and wing vibration strategies, are more successful pollinators than other bees. Some bumblebee species are sold mainly for use in greenhouses. You buy a hive of bees or a box and put it in your greenhouse. Generally, to ensure their longevity, you have a supplementary food supply, such as pre-packaged nectar. Because of the greenhouse's closed setting, it is necessary not to use pesticides on plants that kill bees.

  To keep plants, dry, greenhouse fans circulate air, but they usually do not produce enough airflow to pollinate the flowers. The release of pollen is influenced by temperature and humidity, so pollination attempts are insufficient without the correct environmental conditions. Within greenhouses, high levels of humidity trigger pollen to bind together in clumps, inhibiting the proper dispersal of individual grains. You may use fans to dry plants before pollinating them to solve this.

- *In-Ground vs Pots:* 99 percent of people using a greenhouse would plant them in pots so they can pass them around. Because it will have lesser leaves, the tree will produce less fruit compared to if it were on the ground. However, this does not affect the size of its fruit. Flexibility is the problem with trees grown in the soil inside a greenhouse. After a couple of years, you will want to reorganize your shelves and beds and the tree will do better in a different section. If it is rooted in the ground, this is almost impossible or a lot more effort. Weeds can become another concern. The depth of the roots would make a full layer of weed protection difficult to add. One has got to leave a hole in the tree. This gives weeds an opportunity to begin to spread.

When one chose to grow fruit trees in containers, there are some things that need to be considered:

- *Pot choice:* Container dimensions will depend on the root system of the fruit tree chosen. Fruit trees with tracer roots exhaust the soil around the surface with water and nutrients. Therefore, it is easier to use a pot that is broader than deep. Deep in the soil, fruit trees with tap roots will search for their needs: they need a pot far deeper than a broad one. After a few years, you will have to plant these in the ground. Fruit trees that are solid and deeply rooted should be potted in a wide jar. Bear in mind that the deeper and wider the pot, the more room the roots need to grow strongly. As the thin plastic ones can easily break, it is advisable to use clay pots. Also, to ensure the stability of the tree as it gains height, it is desirable to use a stronger component. Start with a pot of at least 16 inches in all dimensions, if you are planting a young fruit tree. After a couple of years, you will have to re-pot if you begin with the smaller suggested scale. You do not need repotting for the first 5 years if you want a larger one immediately (20+ inches).

- *Soil and Compost:* High-quality soil that is high in nutrients and not polluted with pests is still used. Select a blend of leaf compost (3/4) and topsoil (or garden soil). After many years, this layer depletes, being devoid of any nutrients. Hence, fertilizers will be the only nutrients that nourish the fruit tree.
- *Fertilizing:* For fertilization, compost and earthworm humus are the best choices.

Figure 9.3: Hand watering is deemed uneconomical because it is costly for labor, yet it is relatively inexpensive for automated systems. In addition, this job is often left to the least skilled employees and most automated systems will easily re-pay the economic effect of sloppy watering on the quality of the plants.

Source: https://www.123rf.com/photo_95929965_greenhouse-watering-system-in-action.html

- *Watering:* Another main point in the treatment of every plant is irrigation. The need for regular watering is one drawback to growing in pots. The soil dries out even more easily than the soil on the field does. The roots fetch water deeper into the soil in the open field, but this is not the case in pots. This allows the plants to dehydrate more readily in pots. During their growing season, fruit trees like cool ground. Installing a drip irrigation system is advisable if you are unable to do it physically. This will save them from going through even intermittent droughts. Momentary droughts can hold back their development or even stop it. Each fruit tree has its own water needs, take note of this when your drip system's timer and drip gauge are set. When the top 1-2 inches of soil is dry, water and pour water until the bottom of the container drains. They can be watered less often as plants grow more extensive roots. The type of water is another significant factor. Rainwater, or water without lime, is most suitable for fruit trees.
- *Exchanging the soil and repotting:* Users need to adjust the soil every year or two to deliver the best nutrients to the plant. With each repotting, if the subject is still young, increase the pot's diameter: normally 1 inch per year. After a while, due to the weight and scale of the pot, repotting would become more difficult. Users can then stick to only altering the surface. This includes removing rich soil with the first 2 inches of soil (manure soil, compost with new soil). Users do need to take the tree out of its pot every 4 to 5 years and replace part of the soil. To trim some of the roots, users should take advantage of this tedious process if required.
- *Trimming roots:* This procedure is conducted only on topics that have reached maturity. It requires taking the root ball out from the pot, removing part of the root for the fine roots with pruning shears and using a saw for larger ones. Users must make sure that every cutting instrument they use is disinfected. Farmers can pick and cut 2-4 inches of the roots from an area that is around 1/4-1/3 of the root ball. They could not replicate the procedure on another region of the root ball for another two years. To add new and rich substrates to the bottom of the pot, farmers should take advantage of this new "space saving".
- *Temperature:* One has more influence over a greenhouse's environmental conditions. Depending on the type of fruit tree one

has, it is possible to regulate the temperature (to an extent). The cold does not tolerate fruits like citrus. Potted plants are more vulnerable to cold and less resistant than if they are grown in the soil. The soil does not play its function as an insulator in a pot anymore. This exposes strong variations in temperature to the strain and roots. To prolong the growing season through the winter months, investment in a gas/electric heater is important. Or an evaporative cooler to extend beyond the summer months of the growing season.

Regarding taking care of the plant, fruit trees normally need less attention than other crops. One can achieve a tastier and more abundant harvest when they take care of the needs of the tree:

- *Trimming and pruning of fruit trees:* The fruit trees go into a vegetative standstill after losing all the leaves (in winter). This is a time for the tree to come to rest. This enables one to carry out detailed work on the tree without damaging it (such as pruning). From the first year on, the pruning of fruit trees can be achieved. Bit by bit, one will have to do this before they become robust and well-developed trees. To ensure good fruit production, pruning helps to regulate the growth of these species. Appropriate sharp and clean instruments should be used so that the pruning impacts the fruit tree as little as possible. They will prevent the spread of fungi through the cut areas in this manner, and the tree will suffer less. The trees may grow taller and taller after a few years. This is when the farmer needs to begin to trim the top. Otherwise, the roof of your greenhouse can break through them, and this is certainly something that one needs to stop.
- *Protection from pests and unsuitable temperatures :* Although the fruit trees can be easily shielded from threads, they must also maintain a specific temperature and look out for pests. Using a gas or electric heater may be a good way to fight frost, as described before. Farmers may have to add a bit of protection from the sun and warm air in summer. There is still a possibility that plants, soil, or tools that are infested with certain pests and diseases can be bought. They might also be taken into the greenhouse by certain beneficial insects. Any indication caused by fungi, aphids, mites, and other pests must always be watched out for. The spread in your greenhouse would be restricted by getting rid of them as rapidly as feasible.

Figure 9.4: In a greenhouse, the most common pests include insects for sap-feeding, pollen feeders, caterpillars, and slugs. Some are considerably more difficult to manage than others, making continuous monitoring crucial for the effective development of greenhouses.

Source: https://www.gardeningknowhow.com/special/greenhouses/greenhouse-plant-pests.htm

- *Choosing a rich fertilizer and soil:* Potted or in-ground fruit trees require rich and fertile soil, as with any planted outdoors. To produce delicious fruits, all fruit trees feed on several minerals and nutrients. That is why they need soil that is extraordinarily rich and fertile. For a farmer, the addition of fertilizers would be a daily activity. Natural fertilizers such as earthworm humus, guano, or manure are better used. With fertilizers, however, one must be careful. As natural as they may be, if we use them in abundance, they may harm our tree roots. Depending on the type of fruit tree and on the choice, the type of soil, and volume of fertilizer varies. Before applying fertilizer, it is advisable to do a little soil research.

  Regarding the watering of the plants, any soil tiles must be sacrificed to position the sprinklers if you want to use automatic sprinklers to water crops. For any sprinkler, the increasing area is too wide to water the center from any side. In addition, only the square types of sprinkler irrigation do not cover the 12-by-10 outline of the farmable field. The several possibilities generate different ideal placements of sprinklers that cover the soil tiles while using the fewest possible sprinklers in plantable areas. On the wooden border, those sprinklers are eventually mounted. With iridium sprinklers, only 4 crop spaces (3.3 percent) are needed, while 12 crop spaces (10 percent) are occupied with quality sprinklers (Charles Neal Stewart, Alisher Touraev, Vitaly Citovsky, & Tzvi Tzfira, 2011).

9.3 FRUITS THAT CAN BE GROWN IN GREENHOUSES

Although tomatoes and strawberries may be about as far as many farmers want to go when it comes to growing fruit under glass, these are not the only viable choices available. A variety of fruits, like grapes, peaches, and oranges, can all be grown in remarkably undemanding ways even in the unheated greenhouse.

- *Grapes and peaches:* Many people believe that cultivating vines inevitably requires high temperatures, but it is completely feasible to have your own grape vine with a little care to pick the right variety. Varieties such as "Buckland Sweetwater" or "Black Hamburgh" are suitable for cool environmental conditions, and the holder of a heated greenhouse has many choices available, of course. Vines need open, free-draining soil to grow, and because water-logged soils can impact them very badly, it is essential to get this aspect of their cultivation correct. As they are grown, young plants will need support-stout canes are the usual option-the new spring growths are trained to support them as they develop. To encourage a successful harvest, vines grow their flowers and then their fruit on the new growth of each year; the side-shoots grown in the summer should be pruned in the coming winter, slicing them back to their last bud, a technique known as "spur" pruning (Cees Sonneveld & Wim Voogt, 2009).

  To keep them in peak condition, the plants would require some care throughout the year and must be watered regularly to ensure that ample water reaches all the roots and fed with a high-quality fertilizer. They need to be thinned as the fruits grow to give each grape the space it requires to expand to a good size—a procedure normally done with long scissors with pointing ends, working up from the bottom of the bunch and extracting the smaller ones. Sadly, vines suffer from a variety of pests, like aphids, red spider mites and scale insects, so at the first sign of trouble, a vigilant look out must be maintained and immediate action taken. The fruits can be susceptible to mildew as well. But with care and a little commitment, a few bunches of home-grown grapes can be produced even by the smallest greenhouse.

For countless years, it has been the tradition of using the greenhouse to produce better and larger crops from varieties of fruit that are typically incredibly happy to grow outdoors. Anywhere that did not boast a peach tree and a grapevine developing under glass, there used to be scarcely a big home. A whole greenhouse will be given over to their production often. These were typically lean-to houses in the walled garden with grapes filling the lower roof portion and fan-trained to the wall among apricots, nectarines, and peaches. In some of our older and larger country houses, this can still be observed, but each year the sight is becoming rarer. Many have been replaced by crops such as tomatoes that can be sold to offset some of the ever-mounting expenses (Charles Neal Stewart, Alisher Touraev, Vitaly Citovsky, & Tzvi Tzfira, 2011).

However, grapes and peaches are still common, and there is no excuse why they should not be cultivated in any greenhouse but the smallest. You would need an area of at least 6ft square (4sq.m) to house a fan-trained peach tree, to give you an idea. Currently, grapes are commonly cultivated outside the greenhouse before ever being trained on the roof and afterward led through a hole in the wall. This is not the best method because, before the roots, the top growth begins to expand in the spring, but it leaves the greenhouse border open for other crops, such as tomatoes. However, the use of growing bags for tomatoes is so much simpler, and it suggests that the vine can be grown indoors. Grapes are not too fussy for the type of soil, but they must be drained freely and contain plenty of organic material. A great way to do this is to start preparing a 3ft square (1sq.m) and 3ft (1m) deep planting position. The best time for planting is early spring, but autumn and mild periods during the winter are very appropriate.

Often try to purchase a two-year-old plant that grows in a pot. This is cultivated with the roots pushed out of the rootball and if there is more than one, only the best shoot is kept. This is drastically cut to a bud about 1ft (30cm) from the ground or to the first two buds inside the greenhouse, if grown outside and brought in. Black Hamburgh is arguably the best black grape in Britain for growing under glass and it has outstanding consistency. For peaches, which should be planted in the greenhouse and afterward trained up a wall, we cannot follow the same routine. This makes it necessary for a lean-to home. With a regular-shaped home, growing the vine in a pot is easier. Although other good varieties exist, for all-round results, Peregrine peach and Lord Napier nectarine are difficult to beat. Train the tree as a fan; this implies you will need 6 in (15 cm) horizontal training wires or two separate brick courses on the wall. A fan-trained tree must never be ignored, which means that during the summer quite a lot of work will be required, such as nipping out unnecessary shoots, yanking back and tying in preserved ones and two thinning sessions of the fruitlets. If you think this is going to be too much to deal with, the solution might instead be to grow the tree in a pot (Cees Sonneveld & Wim Voogt, 2009).

- *Nectarines*: It is possible to grow nectarines in either unheated or cool greenhouses. As with vines, in good fruit production, choosing the right variety plays a major role. The "Peregrine" peach is an ideal candidate, typically begins to ripen in August and has the benefit of being self-fertile; by comparison, "Hale's Early" requires another variety close by, albeit an excellent choice, if pollination is to be successful. For the cool or unheated greenhouse, "Pine Apple" and "Humboldt" are highly suitable among the nectarines, maturing later in the season, yielding fruits in September and sometimes into October. It takes a little time and effort to keep these trees in good shape, but they will surely pay for it in the long run. The fruit is grown on the wood of the previous year, unlike vines, and it is common to prune and train the tree to retain a fan shape that provides sufficient space for the fruits to grow and makes it easier to select them.

Figure 9.5: In agriculture, there are four main controlled pollinators utilized, with different tendencies for use in enclosed structures. Ultraviolet light can be used and used by all bees to orient themselves towards their world.

Source: http://www.xinhuanet.com/english/2017-05/08/c_136264103.htm

Nectarines produced in the greenhouse tend to have very compact roots and must therefore be watered regularly and fed once the tree has been formed during the growing season. Despite the absence of pollinating insects in the greenhouse, with a small, fine brush, the plants need some support, the usual technique being to move pollen from flower to flower. This needs to be done during their flowering every day, since not all pollen blooms at the same time, and the greenhouse humidity should also be increased to have the greatest chance of winning. It will often be necessary to thin them as the fruits subsequently grow. When the fruits are about the size of walnuts, this work should be finished, thinning out to leave about two fruits to grow per foot of the branch (FAO, 2018).

- *Citrus fruits:* Perhaps quite shockingly, in the greenhouse, tangerines, oranges, and lemons, can all be grown. For the pips to germinate, they only need a temperature of about 13 degrees C (55 degrees F) and will endure the winter comfortably, provided their atmosphere remains above 4 degrees C (40 degrees F). While growing them from pips is the simplest way to raise your own tree, it is not necessarily a quick process, requiring up to ten years for them to start producing their own fruit. Taking adequate cuttings should enable fruiting plants to be produced significantly more quickly if faster propagation appeals. In the greenhouse, growing fruit is not without its difficulties. Having the requirements exactly right, avoiding the diseases and pest ravages, even properly growing the plants, all take a bit of work to accomplish, but in the long run, there is not much to beat being able to walk out to the garden and select your own.

Regrettably, to be efficient, genuine exotics require incredibly special conditions and these can be awfully expensive for the ordinary farmer to achieve, particularly when it comes to heating. However, citrus fruits are straightforward to grow given that the frost can be held out and that only a little extra warmth is sufficient. There is no excuse why pips should not be produced, but you face the same risk as apples and pears; that is, it is doubtful that the offspring will be anything like the parent. For that kind of reason, rather than taking a gamble, it is always easier to buy young plants from a specialist nursery.

If you are planning to grow trees at the edge of the greenhouse, you must first ensure that the soil is well-drained and slightly acidic. Very few fruits on alkaline soil are a success. A good alternative is John Innes Potting Compost No. 3 for pot cultivation. A compost based on soil like this is superior to a compost-based on peat because it provides the plant with much greater stability and the loam content serves as a buffer against unexpected shifts in temperature and water content. Oranges will still be very okay if only a minimum amount of heat is sufficient, but the temperature should not be allowed to drop to zero, and the plants will only need to be kept dry on the humid side of the bone in the winter. The development will begin in March/April and in the summer, the plants usually bloom. This leads to over-wintering in a limited and green condition of the fruitlets, after which in the summer they will develop on and mature. In February, a minimum of 45-50 ° F (7-10 ° C) will bring them into flower for those growers with a little more heat at their fingertips, which will give rise to ripe fruit in the time from October to January. For both schemes, there is no excuse why they should not be placed outside in a sunny and sheltered location from June to August if the trees are growing in pots.

- *Watermelon:* Growing watermelons in a greenhouse has specific steps. Growers are expected to stuff potting mix in a two-inch tray and water the mixture. This should then be placed in a warm area with temperatures between 85 degrees Fahrenheit to 90 degrees Fahrenheit for 24 hours prior to cultivation. The seeds should not be overwatered. For two days, the temperature should be 90 degrees Fahrenheit. After the appearance of the seedlings, the grower can reduce the temperature to 70-to-80-degree Fahrenheit during the day, and 65 to 70 degrees Fahrenheit at night, for them to develop. Immediately the first true leaf emerges, the farmer should fertilize. The second fertilizer should be applied after the second leaf appears. Watermelon seedlings should be hardened off through a gradual decrease in temperature, and reduction in watering and fertilizer. Inside the greenhouse, 4 to 6 feet of space should be created. Thereafter, the seedlings should be transplanted. The grower should create or buy, for your growing watermelons, a stable trellis, and as the fruit matures, make slings. As soon as the developing runners appear, the plants can be fertilized again.

  When your watermelons are trellised, pruning is necessary. It is implemented to resolve a balance between fruit-bearing and vine growth. The average fruit weight is also increased, and the volume of cull fruit is decreased at the same time. As it grows larger on the trellis, watermelons need help. Trellising encourages pollination during the harvest season and decreases the damage to the plants. Trellising encourages pollination during the harvest season and decreases the damage to the plants. They slowly lift the trellis and will require training to hold the growth upright. Small mesh bags or clothes may be used to hold the watermelons as slings. As soon as you can harvest, the bag can easily be removed from the trellis.

  Mulching will improve the retention of moisture in your watermelon, avoid the growth of weeds and eventually combine nitrogen-rich organic material into the soil. Apply a three to four-inch thickness of hay, shredded newspaper, or grass cuttings over your plants. Avoid using any fertilizer rich in nitrogen until it has started producing fruits. Rather than nourishing the fruit, the excess nitrogen would simply contribute to excess leaves and vines.

  Watermelons grow uniformly. In groups, you can harvest the bulk of it. From 45 to 60 days after flowering, they are available to select. When maturing, they do not move from the vine or emit a scent. Other signs of preparation include increased waxiness of the coating. You may also note any withering of the tendril closest to the fruit. When you knock on a watermelon, you can hear a muffled noise. Clip them from the vine, ideally, and ensure that you leave around 1 inch of the stem. Since it has more juice within it, watermelons can appear heavier than other watermelons of similar size.

Figure 9.6: In a greenhouse, growing watermelons allows you to monitor the amount of temperature, light and humidity required for germination and crop production. If the watermelons are big enough, transplant them into your garden or, with the aid of a trellis, keep growing them in the greenhouse.

Source: https://www.youtube.com/watch?v=ZAUDsQQ3qmw

9.4 ADVANTAGES OF USING GREENHOUSES FOR FRUIT PRODUCTION

The collection of produce in a greenhouse has resulted in a range of benefits for farmers that we have described below:

- *Off-season fruits:* Closed processing with the ability to regulate the weather (temperature and humidity) within allows the cultivation of challenging plants, as far as the weather is concerned, in periods in which the environment is detrimental to the growth of the plantation.
- *Enables each crop and stage to decide the most appropriate climatic conditions:* The ability to control climatic conditions helps us to maintain plants at an optimum ambient temperature

and relative humidity, preferring a decrease in the outdoor production time, early yields, and the potential expansion of production campaigns in a year.

- *Better quality and bigger production:* As the plant may not waste energy from its metabolism to cause physiological defenses to protect itself by using this energy in growth and development, helping to increase the quantity and quality of the crop, the prevention of crops is under pressure due to high temperatures and harmful humidities increasing production.
- *Better control of diseases, weeds, and pests:* Growth in a closed environment with favorable climatic conditions for growing can be inferred a priority by the further production of outdoor pests and diseases, but the great benefit you have is that the impact of pesticides is better, since there are no losses resulting from applications, enabling lower doses of plant protection products to be used in applications.
- *Better control and/or protection against weather conditions that are extreme:* The consequences of severe weather conditions, like heavy rain, snow or hail, and series of strong wind gusts do not impact plants by growing within a cover, because the design and pollution covers help poor conditions, protecting the plantations inside. These structures protect plants from the consequences of unseasonal temperature changes, as described earlier (such as an exceptionally hot day in October or frosty morning in May). If left unprotected, even powerful winds or beaming sun rays can have a negative effect on fragile plants. Sunlight is diffused under a transparent cover (as seen in greenhouse alternatives from ShelterLogic) while ample airflow flows throughout. To securely protect your plants and output from the elements, use a greenhouse.
- *Efficiency in the use of productive resources:* It is possible to reduce the use of water and fertilizers by using techniques such as fertigation and hydroponics, as crops invest only what they require for their growth, reducing production costs.
- *Makes it easier to perform crop tasks in a safe and comfortable manner:* The greenhouse framework also offers security for the performance of the requisite cultivation tasks since programming activities can be met without obstruction to climate time.

- *Defense against livestock and invasive plant species:* Plants grown in greenhouses seem to have the benefit of protecting plants from predators and from invasive plant species as well. These plants will also grow optimally since they are not at risk of being consumed by animals. Their natural living standards are also not altered by invasive species that can compete with them and take away their habitat, sunlight, and nutrients. Plants that are grown in greenhouses are also usually less vulnerable to environmental influences, and because of that, the yields would likely be higher.

9.5 DISADVANTAGES OF USING GREENHOUSES IN FRUIT PRODUCTION

The best way to grow year-round by controlling the seasons, especially in cooler climates, is to have a greenhouse. In winter, consuming homegrown tomatoes and starting bedding plants early helps you to do gardening outside of seasonal limits, but greenhouse technology has drawbacks. A greenhouse's regulated climate carries expenses involved with upkeep and pest management. Taking these and other obstacles into account lets you decide whether a greenhouse is the best match for your gardening needs.

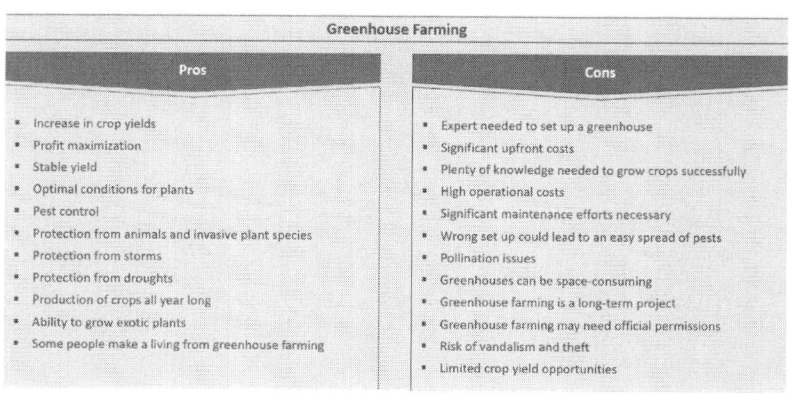

Figure 9.7: Above are the common advantages and disadvantages of greenhouse fruit production. One benefit of greenhouse farming is that the yield during the year is very stable. Plants are properly covered in greenhouses compared to outdoor farming, where the entire yield could be lost due to harsh climatic conditions.

Source: https://environmental-conscience.com/greenhouses-pros-cons/

- *Site requirement for greenhouses:* This primary downside of building one is also discovered by careful consideration of a suitable location for your greenhouse. A greenhouse, especially in winter, should be positioned where it catches the full sun. From November to February, the Clemson Cooperative Extension states that areas on the east side of your property receive optimal sunlight. During warm seasons, deciduous trees shield greenhouse plants from harsh sun, and they also allow more sun after they shed their foliage to reach plants in cool seasons. Electricity and water supplies will be required, which could add to your start-up and maintenance costs.
- *Management of temperature and ventilation:* A greenhouse also becomes a heat trap during warm weather, built to keep in the warmth during cold seasons. Plants may be burned to a crisp if temperatures reach safe levels. Plants freeze to death in cold seasons if temperatures fall too low. According to the Vegetable Research and Knowledge Center of the University of California, this fine balance of temperature control involves heating and ventilating systems, which are the costliest operating costs of a greenhouse. While night temperatures between 50- and 70-degrees Fahrenheit are needed for most greenhouse plants, optimal growth temperatures are plant-specific. This implies that in the same greenhouse, you will not be able to produce a variety of plants, which is one of the drawbacks of greenhouse farming.
- *Issues with pests and diseases:* Pests and diseases also maintain the same shield that a greenhouse offers to keep out unfavorable weather. Trialeurodes vaporariorum (Greenhouse Whiteflies) are key pests that rely on sap and invade more than 250 kinds of plants. Owing to their tolerance to most pesticides, they can be hard to control. Whiteflies proceed through seven life phases prior to reaching maturity. Insecticides that kill nymphs and adults do not respond to eggs and pupae. Neem and horticultural oils control immature stages and regulate adults with pyrethroid insecticides. The humid greenhouse conditions preferred by many plants often foster diseases, which is another explanation for an appropriate ventilation system to be built.
- *Issues with pollination in greenhouses:* Some greenhouse crops need to be pollinated before they can bear fruit, such as tomatoes. Greenhouse fans, which circulate air and hold the leaves dry,

usually do not produce adequate wind for pollination, one of the drawbacks of greenhouse technology. When you cannot fit it into your schedule, manually manipulating flowers to unleash pollen grains is a challenge. To speed up this job, you can use battery-operated pollinating tools, but they still need an operator. As greenhouse pollinators, some bumblebees are bred specifically. The cost of using these bees involves the bees' box or hive and the premixed nectar as an additional source of food for them. Bees can be killed by pesticides that control other insects and diseases, so recognize this when determining how to kill pests.

- *A great deal of knowledge is required to successfully grow crops:* Although many people think it's extremely easy to grow plants, the contrary is happening. When it comes to plant growth in an optimal way, there are so many variables and only one error can lead to the spread of pests or to substantially lower crop yields. To make your greenhouse business effective, plenty of knowledge about farming is therefore needed. Yes, of course, you can make more mistakes and learn from them if you just want to have a little greenhouse for fun. If you expect your greenhouse to be your primary source of income, however, you must ask yourself whether your information would be adequate to prevent potential undesirable economic repercussions.

Chapter 10

Organic Production of Fruit Crops

CONTENTS

10.1 INTRODUCTION

The cultivation of organic fruit effectively avoids the use of several inputs, most importantly synthetic pesticides, and fertilizers, involved with modern farming. Organic farming systems depend, to the greatest extent possible, on control of biological pests, mineral-bearing rock powders, off-farm organic waste, green manures, legumes, animal manures, crop residues, and crop rotations. Such components preserve the fertility of the soil and tilth, provide nutrients for plants, and help manage weeds, insects, and other pests.

Figure 10.1: Organic farming has evolved out of the deliberate efforts of motivated individuals to establish the best possible relationship between men and the planet.

Source: http://ukrsadprom.org/eng/2018/06/20/growing-of-organic-fruit-in-ukraine-and-in-the-world/

10.2 CULTURAL PRACTICES

- *Site selection:* For organic operations, all considerations concerning site suitability for traditional fruit plantings (air and water drainage, etc.) matter even more and more. Although traditional farmers can depend on chemical fertilizers and pesticides to account for poor site choices, such luxuries are largely given up by organic farmers. The organic farmer is also especially concerned about the existence of certain weeds and forage plants. For growers, Bermudagrass, Johnsongrass and some other species can be very troublesome and are challenging to manage by non-chemical methods (Pradeepkumar, 2008).

For farmers involved in organic farming, growing fruit crops provides some advantages. The soil does not need additional tillage or cultivation besides that necessary at the settlement because fruit plantings are perennial, thereby reducing soil erosion. Since the capacity for erosion is low, fruit crops can be produced comfortably and successfully on mountainsides and other sites unfit for tillage farming.

- *Selection of crops:* Environmental constraints (climate, the prevalence of some diseases and pests, suitable soils, etc.) can have a significant effect on the appropriateness of the site or even the bioregion for the organic cultivation of the fruit crop in question. Broadly speaking, it is easier to organically produce certain perennial fruits than others. For example, tiny fruits (berries) appear to be simpler to produce organically in nearly all places than tree fruits. Finally, good growing of organic fruit will largely rely on whether the venture is for domestic production or consumer operations.
- *Preparation on the site:* Attention to the site planning information will go a long way towards minimizing problems with weeds and diseases and ensuring vital planting by enhancement of the soil. For good production, fruit crops do not need extremely fertile soils. Highly fertile, nitrogen-rich soils can potentially encourage too much vegetative growth at the cost of fruiting. The basics of an organic fertility management strategy for fruits are nutritionally healthy soil, sufficient oil pH and abundant quantities of organic matter.

Usually, pre-plant soil enhancement for organic fruit plantings is achieved by a mixture of cover cropping or green manuring and the application of imported materials, including organic waste, rock powders, compost, and manures. Applying the necessary quantities of lime or sulfur is of key significance at this point. Depending on the crop to be cultivated and the outcomes of soil tests, the need for sulfur or lime can be known. It is also easier to change the oil pH with lime (to boost the pH) or sulfur (to lower the pH) prior to planting. Many fruit plants, while they tolerate a pH range around 5.5 and 7.2, function better around pH 6.5. An exemption is blueberries. Preferably, pH 4.8 to 5.2 soil testing can also be utilized to direct treatments of manures and other rock powders to prevent nutrient imbalances. They need acid oil.

In addition to contributing to soil fertility, cover crops and green manures may be part of an effective pre-plant weed suppression plan as smother crops. The fundamental strategy starts with plowing under sod or other established vegetation, planting a cover crop to prevent the growth of weeds, tilling, and establishing plants under the cover crop. Until setting up seeds, many rotations of cover crops are also used (Pradeepkumar, 2008).

Based on variables such as seed cost, available equipment, soil erosion potential, type of soil, and seasonal rainfall, the range of specific cover crops and their maintenance differs with the region. For example, crotalaria at different centers possibly progressed less effectively than sesbania since it is poorly suited to thin, wet soils. Aggressive maize varieties and forage soybean varieties have several other warm seasons cover crops that may be included. To improve drainage and to minimize root rot disease issues, crops like berries (strawberries) are also cultivated on raised beds.

- *Soil solarization:* Soil solarization is another method for site preparation. Through the hot summer months, the procedure entails putting clear polyethylene plastic on saturated soil to raise soil temperatures to quantities lethal to multiple pests. Solarization removes weeds and prevents several possible issues with soil diseases and nematodes. Soil solarization is rather costly and is not commonly used to prepare sites for perennial fruits, with the likely exception of strawberries.
- *Mulching and Floor Management :* There is a range of ways to maintain the floor or inter-row areas of perennial fruit plantings, varying from clean cultivation to diverse blends of cover crops and organic mulches. A system that ends in full or close-to-full ground coverage is better in aspects of managing erosion at least. If the determination to plant an inter-row ground cover is taken, it may consist of a species suitable to the area and to the maintenance plan of the fruit farmer. Orchard grass and other cool-

season grasses are also preferred when modified because they go latent throughout the summer sun, thus reducing competition for water with the fruit crop. These grasses can also supply abundant material for mulch with adequate fertility management. Many legumes in the warm season are deeply rooted and start competing for water with the trees. Underneath the tree canopy, they must not be allowed to grow. However, fruit trees or vines can be supplied with significant nitrogen by legumes and mulch made from leguminous soil covers. In addition, since they exhaust the water with the fruit plants, legumes (like the grasses) improve the soil's water absorption. Raising the organic matter composition of the soil often improves the retentiveness of the soil's moisture (Konvalina, 2016).

Figure 10.2: One of the most efficient methods of conserving moisture throughout dry periods is the usage of mulch. Since it has a variety of other beneficial impacts, including in non-drought conditions, it should be considered a valuable activity. Many mulch materials are available, but below are mainly addressed organic mulches because they are very productive and can be integrated into the soil to enhance it.

Source: https://eorganic.org/node/4870

In early summer, subterranean clover reseeds itself and falls back in the hottest part of the growing season, creating a relatively dense duff or mulch that suppresses weed. In peach and apple orchards and for several orchard crops, this method has shown validity. Subterranean clovers are not suited to climates whereby winter temperatures drop below 0 °F daily. The relationship involving orchard floor management and the control of pests has also been studied by several researchers. A diverse mix of cover crop spices created to shelter and food for several beneficial species in some area apple orchards or peach orchards, resulting in a decline in orchard pests. Dwarf sorghum, buckwheat, mustards, and various members of the Umbelliferae and Compositae families, without causing too many pests, host large numbers of beneficial insects (Konvalina, 2016).

Mulching with organic compounds such as sawdust, leaves, and straw may supply substantial weed suppression after a plantation is created. Total suppression may be accomplished if spread thickly enough or supplemented with sheets of paper or cardboard as the bottom layer. Mulch application varies greatly by crop. For example, in strawberries, mulch could protect whole aisles among rows or only be positioned next to the bed by creeping weeds to prevent encroachment. Mulch could be mounted in vineyards and orchards mainly underneath individual trees or vines or across the entire path, preferably stretching to the drip line.

Typically, to decrease damage from voles, mulch should be kept far away from the trunk. In winter, this is particularly important. The possibility of crown rot and similar diseases in prone species is also reduced by holding the mulch 8-12 inches away from the trunk, particularly on certain rootstocks. In mulberries, with additional sawdust placed along with the canes, sometimes to a depth of 8 or more inches, sawdust mulch is typically spread along the whole row. Mulching with organic materials enhances the soil by improving water availability and soil aggregation, as well as controlling weeds (Konvalina, 2016).

- *Varietal selection:* Since the plants are perennial and constitute a significant investment in both money and time, it is essential for the location and markets to begin fruit planting with the optimum strains. There is good data accessible from Cooperative Extension, nurseries, and local commercial farmers on varietal distribution.

It is also necessary to acquire clean stock for planting. It is suggested to purchase from trustworthy nurseries that have stock certified to be clear of diseases and insect pests by accredited inspectors. Inheritable characteristics in plants that prevent disease and pest damage are referred to as genetic resistance. For organic farmers, the selection of genetically resistant crops is an especially important control measure, particularly about disease management. Cultivar resistance is the strongest or only control in certain situations, such as a bacterial spot in peaches. Aphid-resistant berries, grape rootstocks resistant to Phylloxera, wooly aphid-resistant apple rootstocks, strawberry cultivars resistant to mites, and peach rootstocks resistant to nematodes are accessible. There is no cultivar of any fruit species with several insect pest resistance as significant as this resistance; thus, means other than resistance would most probably have to be used to defend fruit plants from a network of several pest species (Konvalina, 2016).

10.3 ORGANIC FERTILIZATION PRACTICES

Figure 10.3: Organic gardening stresses the cultivation of your garden so that the soil, plants, and beneficial insects are enriched. This is accomplished by restricting the use of pesticides and synthetic fertilizers and by utilizing products that cultivate the soil of your garden and the ecosystems in it.

Source: https://www.pennington.com/all-products/fertilizer/resources/what-is-organic-fertilizer

In contrast to other crops, harvested fruit, which is mainly water, extracts relatively little nutrients from the soil. A large amount of the fertility requirements of fruit crops can then be met by covering crop management and organic mulches in their use systems and by adding lime and other slow-release rock powders at the pre-plant level. Still, for optimum growth and development, supplementary fertilization is often needed. Any of the knowledge useful for preparing a perennial fruit supplementary fertilization program includes:

- In general, organic farmers use relatively insoluble fertilizer ingredients, such as manures, compost, by-products of plants such as animal by-products such as feather or blood meals or cottonseed meal. Early spring implementation is encouraged to ensure sufficient decomposition and timely nutrient distribution of these slowly available materials. The tendency towards late-season development, which can result in winter damage, is also reduced by an early implementation.
- Surface application without the inclusion of organic fertilizer materials is often wasted by the nitrogen found in those materials. This is particularly true of manure. Inclusion by tillage, however, may harm the roots of the fruit plants and raise erosion.
- Many organic fruit farmers prefer soluble organic fertilizers like soluble fish powder, fish emulsion, or water-soluble blood meal to circumvent some of the 'issues' connected to slowly available organic products, in some cases applying them as a foliar spray. The cautious farmer rarely relies solely on such products, as these are relatively costly.
- As the main part, most organic fertilization programs concentrate on supplementing nitrogen because it is required for the crop in the largest amount. One way to assess the correct rate of application of an organic fertilizer is to obtain a traditional chemical planting recommendation and measure the estimated quantity of organic fertilizer required to fulfill it.
- In addition, the farmer requires to credit the approximate impacts produced by legume cover crops and or leguminous mulches, where they are used, when making fertilizer estimates based on nitrogen. In a living mulch system, a cover crop of subterranean clover, correctly fertilized and inoculated, can carry 200 to 1000 kg of nitrogen per acre annually.

- Whenever the fertilizers overall are imbalanced, basing application rates exclusively on nitrogen content can create issues. For example, repetitive use of poultry manure, which is extremely rich in phosphate, can lead to problems with pollution and zinc deficiencies in the crop. By daily soil testing and modifying fertilizer selection and prices accordingly, these issues can be averted.
- Combining field observations with soil and tissue testing is the most accurate means for evaluating whether fertilization is sufficient. Weak yields, irregular leaf coloration, and poor growth of plants are all signs of a potential nutritional imbalance or deficiency. Nitrogen deficiency is possibly demonstrated, for instance, by less than 10 inches of annual elongation of the branches of most fruit trees. A corky bark may suggest an over-accessibility of manganese in the soil for certain apple varieties.
- A foliar analysis tests the nutrient composition of the leaves and, well in advance of visible signs, may detect a deficiency or excess. Since foliar analysis is a measurement of what the plant is currently taking up, it is more useful than a soil test, whereas soil analysis only tests what is in the soil that may or may not be accessible to the plant. The best guide for modifying supplementary nitrogen fertilization is usually given by annual foliar studies.
- *Organic weed management:* Studies suggest that crop yields and plant vigor will be massively diminished without some sort of weed control in fruit planting. In organic farming, weed control for perennial fruit crops is just one purpose of a weed management scheme. A successful organic weed control plan should pose a minimum risk of erosion, include a 'platform' for the movement of farm machinery, not adversely affect the management of pests or soil fertility, while at the same time minimizing water and nutrient demand for weeds (Helga Willer, Minou Yussefi, & Neil Sorensen, 2010).
- *Organic mite and insect peat management:* A significant difference between pest control in annual crops and perennial fruit crops is that crop rotation is not an alternative (strawberries and to a lesser extent, brambles, are possible exceptions). The long-term nature of fruit production makes it feasible over time to

build up a population of pests. Alternatively, it is also important to maintain populations of beneficial species in such healthy agricultural conditions.

10.4 PLANT VIGOR AND HEALTH

Although it is often overstated, it is necessary for pest control to maintain the plant in general health and good vigor. This adage is more relevant to indirect pests (those pests feeding on foliage, roots, etc.) for fruit plants than to direct pests (pests that feed on the fruit). A seemingly healthy plum tree, for example, can set up a strong fruit crop, then lose it all to the plum. On the other hand, early in the season, the same tree could suffer substantial defoliation by caterpillars; but if it is in good vigor, it can recover and bounce back easily while still generating a profitable crop that year. There are some instances where overall plant health and independence from stress confer a form of "resistance" to some pests that is not theoretically genetic resistance (Helga Willer, Minou Yussefi, & Neil Sorensen, 2010).

10.5 BIOLOGICAL CONTROL

The usage of living organisms to manage the existence of a pest relates to biological control. The destructive mites *Metaseiulus occidentalis* and *Phytoseiulus persimilis*, which target lady bird beetles, spider mites, and green lacewings feeding on aphids and Trichogramma wasps that parasitize the eggs of several pests, particularly codling moth, are examples of helpful arthropods utilized to manage pests in fruit plantations. Many helpful insects can be bought from commercial insects and introduced into fruit plantations. Even so, control of cover crops and adjacent vegetation is more feasible as an insect refuge to accommodate and maintain beneficial indigenous populations. As a rule, it appears that helpful arthropods are not a full control mechanism for specific fruit pests at least for commercial farmers who have a low damage tolerance for fresh fruit. Typically, extra control measures are needed.

Figure 10.4: Biological Pest Control incorporates into the environment a distinct species that either predates the pest or causes illness.

Source: https://pmgbiology.com/2014/12/01/biological-control-of-pests-a-understanding-for-igcse/amp/

- *Biorational and Organic pesticides:* In general, pesticides authorized for certified organic cultivation are obtained from natural sources, rapidly break down in the field and tend to have a minimal environmental effect. Botanical products from plants, insect growth regulators, treatments with synthetic pheromones that cause mating damage, soaps, oils, minerals like sulfur dust and biological pesticides are examples. The word "biorational pesticide" is utilized to define organically appropriate or synthetic pesticides which have a limited effect on the beneficial and environmental insects (Pradeepkumar, 2008).

  Those that have the least toxic control that can be implemented against a pest are called bio-pesticides. Biological pesticides (biopesticides) are both chemically and biologically permissible. Biopesticides vary in part from biological control because they are developed, labeled, and administered like normal pesticides, but also because there is no substantial reproduction in the field of the species involved. Biopesticides are extremely specific to humans or beneficial insects and do not affect them. In fruit pest control, several biopesticides may be used. They must be used at a particular time in the growth cycle of the pest to be successful.

By removing toxic substances from plants that have pesticidal effects, botanical insecticides are developed. In the ecosystem, they are found naturally, short-lived and do not produce damaging residues. Many botanicals, however, are wide range poisons that affect pests and beneficial compounds alike and are not always the preference of biology. Botanical extracts are widely used by organic producers, who are forbidden from using synthetic pesticides. Specially developed soaps rich in fatty acids are effective against many soft-bodied insects like spider mites, leafhoppers, whiteflies, and aphids. Insecticidal soap penetrates the body of the organism and destroys cells and their membranes' regular work, allowing the substances to spill out.

Utilizing certain woody plants like bushes, grapevines, and fruit trees, with a thin layer of dormant oil suppresses pests such as scales, mites, aphids, and leaf rollers by suffocating wintering adults and larvae. Just before bud break, dormant oils should be applied and should never be intermixed with sulfur because there may be foliage damage. So-called summer oils have a relatively low viscosity and, if label caution is practiced, can be used throughout the growing season to combat mites and aphids avoiding plant damage. Insect pheromones are substances that insects generate to enable them to convey things such as the availability of mates and sexual perception. Typically, they are unique to a particular species or genus of insects. Scientists have discovered most of these pheromones can be synthesized and are commonly used to track the appearance of crop pests or their simple presence. This knowledge is widely used for applications of pesticides over time. New technology also makes it possible to use pheromones for the mating distortion of certain pests. For the oriental fruit moth, codling moth, peach tree borer and grape berry moth, mating disturbance pheromones are visible (Pradeepkumar, 2008).

10.6 ORGANIC DISEASE MANAGEMENT

However, there are certain forms of illnesses that are prevalent in almost all perennial fruit crops in the temperate region. For example, fruit rots are common pathogens due to the relatively soft nature and high sugar content of many of these mature or almost mature fruits. By enabling good air circulation and sunlight infiltration into the inner plant canopy, the organic farmer may help reduce the odds of fruit rot. Sunlight and circulating air allow for the drying of the surfaces of the leaves and fruit, hence reducing fungal and bacterial pathogens. In tree crops, this would imply proper techniques of pruning and preparation. In brambles and berries, plant density reduction helps. It is helpful to prevent rank vine growth in grapes and eliminate leaves that cover fruit clusters. A site that allows for good air circulation should be chosen for all fruit crops.

Root rot and intolerance of poorly drained soils are other issues common to many fruit crops. Many pear rootstocks and some apple rootstocks are relatively tolerant to heavy soils that are poorly drained, but even such crops can succumb to waterlogged environments that persist. Prunus species are very unaccepting to poorly drained soils (cherries, plums, peaches, etc.) and are usually prone to root-rotting microorganisms which common in such soils. Other cultural aids to mitigate disease may include things such as holding plants in good health and vigor, removing plants from the planting site by pruning, roughing out diseased plants, removing alternative hosts or sources of inoculum for the disease species.

- *Organic bactericides and fungicides:* The principal bactericides and fungicides utilized by organic farmers are sulfur and copper compounds, but they have disadvantages. If applied improperly, these materials can cause harm to plants. For certain beneficial insects, mites, and spiders, Sulfur is also deadly and may set the stage for more pest issues. The use of copper fungicides , in the long run, can also lead to harmful levels of copper in the soil. In addition, these fungicides are usually subordinate to synthetic alternatives, and need a repeated application to be utilized on a protective schedule. Research is promising with respect to bio fungicides. As a treatment for gray mold, many formulations of the fungus *Trichoderma harzianium* are already on the market (Botrytis). A treatment for powdery mildew in grapes and a protective agent against tree wound pathogens are other bio fungicides now accessible (Konvalina, 2016).

References

1. Basra, A. (2000). *Plant Growth Regulators in Agriculture and Horticulture: Their Role and Commercial Uses.* Florida: CRC Press, 1-82.
2. C.M. Karssen, L.C. van Loon, & D. Vreugdenhil. (2012). *Progress in Plant Growth Regulation.* Berlin : Springer Science & Business Media, 1-82.
3. Castilla, N. (2013). *Greenhouse Technology and Management.* London: CABI, 11-43.
4. Cees Sonneveld & Wim Voogt. (2009). *Plant Nutrition of Greenhouse Crops.* Berlin: Springer Science & Business Media, 1-98.
5. Charles Neal Stewart, Alisher Touraev, Vitaly Citovsky, & Tzvi Tzfira. (2011). *Plant Transformation Technologies.* New Jersey: John Wiley & Sons, 76-86.
6. Davies, P. (2013). *Plant Hormones: Physiology, Biochemistry and Molecular Biology.* Berlin: Springer Science & Business Media, 13-58.
7. FAO. (2018). *Good agricultural practices for greenhouse vegetable production in the South East European countries: Principles for sustainable intensification of smallholder farms.* London: Food & Agriculture Org, pp449.
8. FAO. (2020). *Crop production manual: A guide to fruit and vegetable production in the Federated States of Micronesia.* London : Food & Agriculture Org, 5-69.
9. Frank Kempken &, Christian Jung. (2009). *Genetic Modification*

of Plants: Agriculture, Horticulture and Forestry. Berlin: Springer Science & Business Media, 2-103.

10. Hancock, J. F. (2008). *Temperate Fruit Crop Breeding: Germplasm to Genomics.* Berlin: Springer Science & Business Media, 1-83.
11. Helga Willer, Minou Yussefi, & Neil Sorensen. (2010). *The World of Organic Agriculture: Statistics and Emerging Trends 2008.* London: Earthscan, 7-101.
12. K. K. Jindal & R. C. Sharma (prof.). (2004). *Recent Trends in Horticulture in the Himalayas: Integrated Development Under the Mission Mode.* Delhi: Indus Publishing, 19-85.
13. Konvalina, P. (2016). *Organic Farming: A Promising Way of Food Production.* London: BoD – Books on Demand, 29-83.
14. Ladaniya, M. (2010). *Citrus Fruit: Biology, Technology and Evaluation.* New York: Academic Press, 67-79.
15. Louise Ferguson & Elizabeth Elliot Grafton-Cardwell. (2014). *Citrus Production Manual.* California: UCANR Publications, 21-61.
16. Md Asaduzzaman & Toshiki Asao. (2019). *Improvement of Quality in Fruits and Vegetables Through Hydroponic Nutrient Management.* Berlin: BoD – Books on Demand, 9-85.
17. Nandwani, D. (2014). *Sustainable Horticultural Systems: Issues, Technology and Innovation.* Berlin: Springer, pp395.
18. NIIR Board. (2005). *Cultivation of Fruits, Vegetables and Floriculture.* Delhi: Niir Project Consultancy Services, 3-98.
19. Peter, K. V. (2009). *Basics Of Horticulture.* New India : New India Publishing, 1-11.
20. Pradeepkumar, T. (2008). *Management of Horticultural Crops.* New India : New India Publishing, pp1028.
21. R.K. Salar, S.K. Gahlawat, P. Siwach, & J.S. Duhan. (2014). *Biotechnology: Prospects and Applications.* Berlin: Springer Science & Business Media, 1-86.
22. Riccardo Gucci & Claudio Cantini. (2000). *Pruning and Training Systems for Modern Olive Growing.* Clayton, Australia: Csiro Publishing, 17-93.
23. Richard E. Litz, Fernando Pliego-Alfaro, & Jose Ignacio Hormaza. (2020). *Biotechnology of Fruit and Nut Crops, 2nd Edition.* London: CABI, 1-65.

24. Robert E. Paull & Odilo Duarte. (2011). *Tropical Fruits, Volume 2.* London: CABI, 1-91.
25. Saleh, M. (2011). *Pruning and Population Adjustment in Relation to Fruit Trees Productivity.* Berlin: GRIN Verlag, pp52.
26. Satyapal Singh & M.R. Choudhary. (2012). *Production Technology of Fruit Crops in Wasteland.* SAn Diego: Scientific Publishers, 1-70.
27. Shri Mohan Jain & P.M. Priyadarshan. (2008). *Breeding Plantation Tree Crops: Tropical Species.* Berlin: Springer Science & Business Media, 3-83.
28. T. Radha & Lila Mathew. (2007). *Fruit Crops.* New India : New India Publishing, 1-79.
29. Valpuesta, V. (2002). *Fruit and Vegetable Biotechnology.* London: Woodhead Publishing, 3-25.
30. Varkulevicius, J. (2010). *Pruning for Flowers and Fruit.* Clayton, Australia: Csiro Publishing, pp224.
31. Zhang, Q. (2017). *Automation in Tree Fruit Production: Principles and Practice.* London: CABI, 1-93.

INDEX